Shailesh More

Aging Degradation and Countermeasures

Shailesh More

Aging Degradation and Countermeasures

in Deep-submicrometer Analog and Mixed Signal Integrated Circuits

Südwestdeutscher Verlag für Hochschulschriften

Impressum / Imprint
Bibliografische Information der Deutschen Nationalbibliothek: Die Deutsche Nationalbibliothek verzeichnet diese Publikation in der Deutschen Nationalbibliografie; detaillierte bibliografische Daten sind im Internet über http://dnb.d-nb.de abrufbar.
Alle in diesem Buch genannten Marken und Produktnamen unterliegen warenzeichen-, marken- oder patentrechtlichem Schutz bzw. sind Warenzeichen oder eingetragene Warenzeichen der jeweiligen Inhaber. Die Wiedergabe von Marken, Produktnamen, Gebrauchsnamen, Handelsnamen, Warenbezeichnungen u.s.w. in diesem Werk berechtigt auch ohne besondere Kennzeichnung nicht zu der Annahme, dass solche Namen im Sinne der Warenzeichen- und Markenschutzgesetzgebung als frei zu betrachten wären und daher von jedermann benutzt werden dürften.

Bibliographic information published by the Deutsche Nationalbibliothek: The Deutsche Nationalbibliothek lists this publication in the Deutsche Nationalbibliografie; detailed bibliographic data are available in the Internet at http://dnb.d-nb.de.
Any brand names and product names mentioned in this book are subject to trademark, brand or patent protection and are trademarks or registered trademarks of their respective holders. The use of brand names, product names, common names, trade names, product descriptions etc. even without a particular marking in this works is in no way to be construed to mean that such names may be regarded as unrestricted in respect of trademark and brand protection legislation and could thus be used by anyone.

Coverbild / Cover image: www.ingimage.com

Verlag / Publisher:
Südwestdeutscher Verlag für Hochschulschriften
ist ein Imprint der / is a trademark of
AV Akademikerverlag GmbH & Co. KG
Heinrich-Böcking-Str. 6-8, 66121 Saarbrücken, Deutschland / Germany
Email: info@svh-verlag.de

Herstellung: siehe letzte Seite /
Printed at: see last page
ISBN: 978-3-8381-3470-3

Zugl. / Approved by: München, TU, Diss., 2012

Copyright © 2012 AV Akademikerverlag GmbH & Co. KG
Alle Rechte vorbehalten. / All rights reserved. Saarbrücken 2012

Preface

Reliability of analog and mixed signal circuits fabricated using complementary metal oxide semiconductor technologies in the deep-submicrometer technology nodes is significantly affected by process, voltage and temperature (PVT) variations. Degradation induced due to aging mechanisms like bias temperature instability, conducting and non-conducting hot carrier injection in n-channel and p-channel MOSFET devices leads to additional challenges in design of reliable circuits. PVT variations and aging mechanisms together lead to lifetime degradation of device and circuit performance. Introduction of alternative high-κ dielectric stack with metal gate has on one hand improved transistor performance but on the other hand has introduced effects like positive bias temperature instability which degrades the transistor parameters over lifetime. Hence reliable and robust operation of semiconductor integrated circuits over their specified operating lifetime is an important specification target for products fabricated in current and future nano-technology era.

We can no longer gain from the reliability experience achieved from the older materials and technologies. Accurate prediction of aging induced performance degradation is important right from the design phase in order to avoid chip failures at client site and expensive design re-spins. However the new challenges are posed by interaction of different aging mechanisms causing enhancement or slow down of the overall circuit performance degradation. Successful integration of the new technology hence depends on the quick understanding of reliability physics and study of aging degradation behavior beyond the level of single transistors i.e. further extended to circuit domain. These aging effects which contribute to both temporal and permanent parameter shifts in transistors cannot be handled only by process improvements. Mitigation of reliability degradation needs to be addressed by design strategies and techniques in the form of countermeasures in advanced technology nodes.

This book presents the outcome of investigations on the effects of aging mechanisms induced parameter shifts and performance degradation in analog and mixed signal circuits. The lifetime degradation induces threshold voltage and drain current shifts that can result into mismatch in matched transistor pairs which is especially important for analog and mixed signal circuit's accuracy. The investigations are done based on analytical evaluation, aging simulation and measurements using sample circuits implemented in state-of-the-art 32nm high-κ metal gate CMOS technology. Circuit performance degradation due to process variation, variability in aging induced parameter drifts and recovery effects is not treated in this work. Calibration and correction techniques suitable for overcoming time varying aging induced circuit performance degradation are proposed and investigated.

The structure of this book is designed to guide the reader from important aging mechanisms, over to aging degradation in important analog building blocks and countermeasures to overcome these effects, leading to effects of aging induced performance degradation on switch, ring oscillator and analog to digital converter (ADC) circuits.

Chapter 1 gives an overview of the importance of precise analog and mixed signal (AMS) circuits and the challenges it faces under aging degradation. It points the need of accurate prediction of circuit lifetime degradation under aging mechanisms in current nano-technology era. Presenting the state-of-the-art scientific research in the field of AMS circuit reliability, the contribution of this research work is highlighted.

Chapter 2 introduces the high-κ metal gate CMOS technology and related aging wearout mechanisms treated in this work. The modeling of the degradation induced by these aging mechanisms on transistor level is explained using a sub-circuit model. And an aging simulation flow to evaluate the reliability of the circuit post aging is discussed.

Chapter 3 presents an analytical approach to evaluate the contribution of different aging mechanisms to performance degradation of linear circuit. This technique proves to be in good agreement with circuit simulations, but with considerably less computing effort and providing more intuitive insight into the various degradation contributions. The concept of accelerated aging to perform quick circuit lifetime prediction measurements under aging degradation is introduced.

Chapter 4 summarizes the findings related to aging degradation in operational amplifier circuits, both in closed and open loop configurations. The importance of circuit topology selection concerning reliability is explained with an example of comparison between two operational amplifier design implementations viz., simple Miller amplifier and folded cascode amplifier. Based on the methodology explained in chapter 3, contributions of different aging mechanisms towards aging induced performance degradation in these two topologies are compared.

Chapter 5 proposes two solutions viz., chopper stabilization and autozeroing to mitigate the effects of aging induced performance degradation in differential circuits. The concept of using chopper stabilization technique as a degradation countermeasure is proved with measurement results on the test chips implemented using 32nm high-κ, metal gate CMOS technology.

Chapter 6 presents the contribution of different aging mechanisms towards parameter drifts in transistors of the ring oscillator circuit. The increased contribution of hot carrier mechanism in minimum channel length devices toward circuit lifetime degradation is highlighted. And an adaptive bipolar tracking technique to monitor and compensate aging degradation in the ring oscillator circuit is demonstrated by measurements.

Chapter 7 presents the contribution of different aging mechanisms towards parameter drifts in transistors of the transmission gate switches used commonly in switched capacitor circuits. Further, ineffectiveness of using bipolar stress as online countermeasure to compensate aging degradation using accumulation stress is discussed.

Chapter 8 introduces analysis and evaluation of aging degradation in high resolution Nyquist rate successive approximation register (SAR) ADC circuit. The impact of aging on building blocks of SAR ADC viz., input buffer and comparator, and its individual and combined effect on ADC performance is evaluated under asymmetrical input stress condition. The need to implement special countermeasures which can correct time varying errors resulting from stress induced degradation in high resolution ADC's implemented in deep-submicrometer CMOS technology is highlighted.

Chapter 9 presents analysis and evaluation of aging induced performance degradation in oversampling sigma-delta ADC circuit. Investigations are carried out on aging degradation of different building blocks in fully differential third-order, 2-stage, multi-bit (17-level) sigma delta ADC implemented in 32nm high-κ metal gate CMOS technology.

Chapter 10 summarizes conclusions from this work and outlook.

Contents

Preface   i

1 Introduction   1
    1.1 Motivation 1
         1.1.1 Analog and Mixed Signal Circuits 1
         1.1.2 Aging in Deep-submicrometer CMOS Technology 2
         1.1.3 Impact of Aging on Analog and Mixed Signal Circuits 3
    1.2 State-of-the-art 4
    1.3 Contributions of This Work 6
    1.4 Summary 7

2 Transistor Level Modeling of Aging Mechanisms   9
    2.1 High-κ Metal Gate CMOS Technology 9
    2.2 Aging Degradation Mechanisms 10
         2.2.1 Bias Temperature Instability 10
         2.2.2 Hot Carrier Injection 12
    2.3 Transistor Level Modeling 14
    2.4 Reliability Simulation Flow 16
    2.5 Summary 18

3 Circuit Level Analytical Evaluation and Accelerated Aging   19
    3.1 Methodology for Analytical Evaluation 20
         3.1.1 Steps for Analytical Evaluation 20
         3.1.2 Application of Methodology 21
         3.1.3 Insight into Aging Degradation Mechanisms 25
    3.2 Concept of Accelerated Aging 26

3.3	Summary	28

4 Aging in Operational Amplifiers   **31**

4.1	Operational Amplifiers	31
	4.1.1 Closed Loop OTA	32
	4.1.2 Open Loop OTA	33
4.2	Comparison Between Aging of Different OTA Topologies	35
	4.2.1 Simple Miller OTA	35
	4.2.2 Folded Cascode OTA	36
4.3	Summary	41

5 Active Countermeasures against Aging Degradation   **43**

5.1	Chopper Stabilization	44
	5.1.1 Introduction to CHS Technique	44
	5.1.2 Reduction in Aging Degradation using CHS Technique	46
	5.1.3 Measurements	48
5.2	Auto Zeroing	53
	5.2.1 Introduction to AZ Technique	53
	5.2.2 Reduction in Aging Degradation using AZ Technique	55
5.3	Summary	57

6 Aging in Ring Oscillator Circuits   **59**

6.1	Ring Oscillator	59
6.2	Aging Monitor and Compensation Circuit	63
6.3	Measurements	66
6.4	Summary	69

7 Aging in Switches used in Switched Capacitor Circuits   **71**

7.1	Switches in SC Circuit	71
7.2	Aging Monitor for Switch Degradation	74
7.3	Measurements	76
7.4	Countermeasures	79
7.5	Summary	80

8 Aging in Successive Approximation Register ADC   **81**

8.1	Introduction to SAR ADC	81
	8.1.1 SAR ADC Model Implementation	83
8.2	Aging in SAR ADC Building Blocks	84
8.3	Effect of Aging on SAR ADC Performance	85
	8.3.1 Effect of Buffer Aging on SAR ADC	85
	8.3.2 Effect of Comparator Aging on SAR ADC	87
	8.3.3 Combined Effect of Buffer and Comparator Aging on SAR ADC	89
8.4	Countermeasures	91
8.5	Summary	91

9 Aging in Sigma Delta ADC — 93

9.1	Introduction to Sigma Delta ADC	93
	9.1.1 Sigma Delta ADC Implementation	94
9.2	Aging in Sigma Delta ADC Building Blocks	97
	9.2.1 Effect of Integrator Aging on Sigma Delta ADC	97
	9.2.2 Effect of Multi-Bit Quantizer Aging on Sigma Delta ADC	99
	9.2.3 Effect of Current Steering DAC Aging on Sigma Delta ADC	101
	9.2.4 Combined Effect of integrator, Quantizer and DAC Aging on Sigma Delta ADC	103
9.3	Countermeasures	103
9.4	Summary	104

10 Conclusions and Outlook — 105

10.1	Conclusion	105
10.2	Outlook	107
	10.2.1 Variability in Aging Degradation	107
	10.2.2 BTI Recovery Effect	108
	10.2.3 Novel Devices and Design Strategies	108

List of Symbols and Abbreviations — 111

References — 115

Chapter 1

Introduction

This chapter gives an introduction to various topics that will be investigated in this work. It highlights the importance of precise and reliable analog and mixed signal (AMS) circuits in today's electronic system design. The challenges faced by these circuits under aging degradation are discussed. It points out the need of accurate prediction of circuit lifetime degradation under aging mechanisms in current deep-submicrometer technology era. A brief overview of the current state-of-the-art scientific research in the field of AMS circuit aging is presented and the contributions of this research work are introduced.

1.1 Motivation

1.1.1 Analog and Mixed Signal Circuits

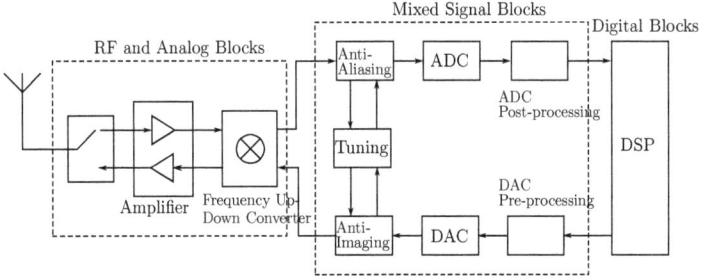

Fig. 1.1: An analog mixed signal integrated circuit

Real world signals such as temperature, speed, pressure, flow, voice are a few examples that we deal in everyday life which are all analog in nature. Analog circuits are found in many electronic systems with human interface applications for example displays, speakers, microphones, cameras, automobiles, sensors, mobile phones, etc. to process analog input signals and generate analog output signals. With scaling being more favorable to digital circuits compared to analog circuits in terms of area, cost and robustness towards noise and variation, more and more functionality is being shifted to digital domain and analog has moved into the background. Due to this paradigm shift mixed signal circuits which serve as connection between analog and digital signal processing domain are indispensable as illustrated in figure 1.1. Precision and accuracy of analog and mixed signal (AMS) circuits is linked to good matching of transistor pairs in structures like current mirrors, differential pairs, etc. and high signal-to-noise ratio which is linked to the available signal swing and voltage headroom.

Non-constant field scaling of CMOS technology has resulted in smaller and faster AMS circuits but with deteriorated linearity and accuracy. In deep-submicrometer technology era with supply voltages around 1V, the downscaling of transistor threshold voltage has de-accelerated due to transistor variability, matching and leakage issues resulting into limited voltage headroom. This limited voltage headroom makes cascoding of transistors difficult to implement. Further due to limited degree of control on device parameters like dopant concentration, channel length and oxide thickness transistor mismatch increases due to the process variations. In addition transistor parameters like threshold voltage and drain current are affected due to aging which results in mismatch and degradation in performance over circuit operating lifetime.

1.1.2 Aging in Deep-submicrometer CMOS Technology

Integrated electronic systems with AMS circuits fabricated using CMOS technology find a wide range of applications ranging from life critical field; aircraft, pacemaker and automotive, consumer electronics field; television systems, mobile, camera and gaming station, to non-critical field; toys and electronic greeting cards. CMOS transistors used in these circuits are expected to degrade (age) with time. This causes the circuit performance to deviate from its specifications measured post fabrication. So device and circuit reliability evaluation is of prime practical importance. The recent CMOS technologies have witnessed slowing down or stopping of supply voltage (V_{DD}) and threshold voltage (V_{th}) scaling because of the non-scalability of sub-threshold slope whereas the transistor gate length (L) and thickness of gate oxide (t_{ox}) is continuing to scale down. This results into a net increase in lateral electric field, effective channel field and the vertical oxide field. Moreover, with scaling of device geometry and the increase in device number, power consumption rises resulting into rise in the operating temperature which produces another big issue with respect to device reliability. The introduction of first nitrogen and then high-κ in the gate oxide stack has lead to enhancement of oxide degradation in both pMOSFET and nMOSFET devices. This leads to enhancement of different aging degradation mechanisms in the integrated circuits fabricated using state-of-the-art deep-submicrometer CMOS technology.

Aging degradation mechanisms can be classified into destructive and non-destructive categories, depending on if it leads to transistor hard failure (e.g. gate-oxide breakdown) or wearout (e.g. bias temperature instability (BTI), conducting, non-conducting hot carrier injection (CHCI, NCHCI)). Hard failures due to destructive stress are completely unacceptable since it can partially or completely disrupt the functionality of the circuit.

Motivation 3

On the other hand wearout mechanisms due to the non-destructive stress are acceptable unto a limit which is defined by the desired circuit accuracy and precision. So to predict if the circuit meets the target lifetime expectation, performance degradation over lifetime under non-destructive stress effects must be analyzed by the designer. Accurate prediction of aging degradation is important to avoid expensive re-spins, for gaining the consumers trust and to correctly define the warranty period and cost of the product. In order to perform quick practical aging predictions to evaluate the lifetime reliability of an integrated circuit, it is necessary to map the end-of-life use case condition of the product (e.g. mobile phone use case of 4 Years, 85°C, 105% of worst case V_{DD}) to an meaningful and accurately mapped accelerated stress condition (e.g. 10^3s, 125°C, 120% of worst case V_{DD}). For these evaluations the AMS circuit is assumed to be well designed and functioning perfectly at time zero. This acceleration allows the stress conditions to shrink the 4 year product life to a 10^3s period so the reliability of the circuit can be studied in laboratory and guaranteed. The shrinking in lifetime of a MOSFET device is possible by elevating the stress temperature, bias voltages and time.

1.1.3 Impact of Aging on Analog and Mixed Signal Circuits

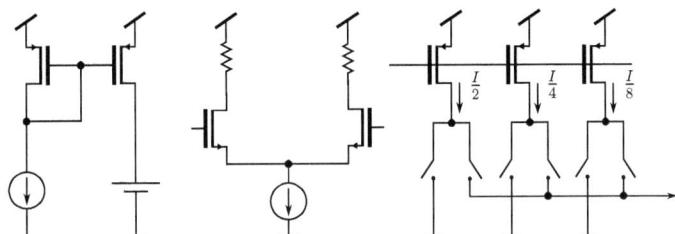

Fig. 1.2: Matching sensitive analog and mixed signal circuits

Current mirror, operational amplifier and bandgap reference circuits are some of the very basic building blocks of AMS systems. The precision and accuracy of these basic building blocks is linked to the matching of the transistor pairs as illustrated in figure 1.2. The reliability performance of all such matched pair circuits depends closely on their aging differential. Analog circuits always witness DC voltages for biasing purposes irrespective of the input signal unlike digital circuits. Further in addition to the applied DC bias voltages, a high temperature may also exist on the chip because of the high transistor density. Thus the failure rate varies as a function of stress voltage, temperature and time. Further the maximum allowed margins of process and aging degradation induced parameter drifts and variations are lower for analog applications and high resolution mixed signal circuits [1].

The transistors in typical AMS circuit are operated either in active mode or power down mode. Operation in either of these modes can induce aging degradation in the transistors depending on the surrounding bias conditions [2]. In the circuit active mode, the transistors are usually operated in saturation region with gate to source overdrive voltage $V_{od} = |V_{gs} - V_{th}|$ of around several 100mV and drain to source voltages $|V_{ds}| > |V_{od}|$. Diode connected transistors are less prone to aging degradation due to their low biasing values with $|V_{gs}| = |V_{ds}|$. Other transistors can see high voltage conditions enough to induce aging degradation due to BTI and/or HCI depending on the input signals and the circuit configuration (closed loop, open loop, feedback, etc.). Asymmetrical input signals lead to aging degradation induced offset voltages in matched differential pairs [3].

In a typical power down mode the bias currents are switched off to avoid power consumption of the inactive circuit, but the supply voltages are not driven down in order to allow for fast reactivation of the circuit. In this case the potentials of the internal nodes are determined by the input signals and the sub-threshold or off state leakage currents of the transistors. All the transistors connected in the current mirror configuration are not prone to aging degradation in this case because the diode connected transistors lead to low gate voltages. The remaining transistors can be affected by BTI stress depending on the input signals.

In case of specific circuits like ring oscillator the transistors see high gate to source voltages ($|V_{gs}|$), being switched between V_{DD} and V_{SS} leading to BTI and NCHCI degradation. Also the transistor here experience high drain to source voltage ($|V_{ds}|$) during signal transition phase resulting into degradation due to CHCI. CMOS transistor switches with bi-directional current flow typically used in switched capacitor circuits experience similar stress conditions like the transistors in the ring oscillator circuit but with lower $|V_{ds}|$ values resulting into low CHCI and NCHCI degradation.

Thus accurate evaluation of aging degradation is required on circuit level to obtain realistic risk evaluation for precise reliability qualification. Simply sizing up devices, such as is done to reduce process variation and HCI effects offers little relief to NBTI and PBTI degradation effects on circuits. The AMS circuit designers need to move one step further to include device aging impact into consideration, so that the circuit can meet the specifications at end-of-life (EoL). Special circuit techniques are needed as countermeasure for these aging degradation effects [4].

1.2 State-of-the-art

A trend of rising interest among the research community in the field of AMS circuit reliability is witnessed in this deep-submicrometer technology era facing severe reliability challenges. The plot in figure 1.3 illustrates the approximate number of publications from academia and industry in IEEE conferences and journals over a period of last 10 years in the field of AMS circuit reliability related to aging degradation mechanisms like NBTI, PBTI, CHCI and NCHCI. A summary of the state-of-the-art and related articles published over the last decade is presented in this section.

A comprehensive overview concerning reliability of MOSFET devices implemented in $0.25\mu m$ technology under analog operation for NBTI and CHCI degradation is given in [2]. In the similar direction an overview of analog circuit reliability for an advanced

State-of-the-art

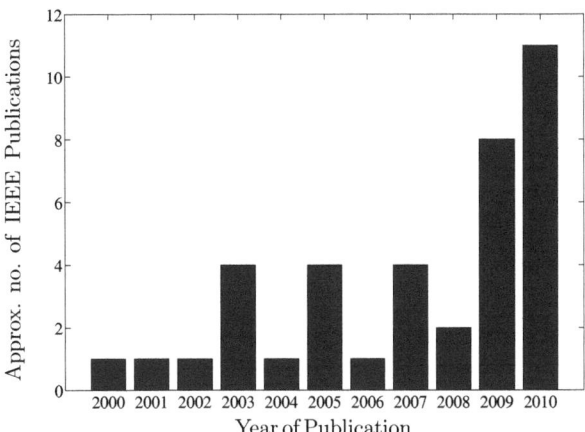

Fig. 1.3: IEEE publications in the field of analog and mixed signal circuit reliability targeting BTI and HCI mechanisms

32nm high-κ metal gate technology is presented in [5]. Physical design and reliability issues in deep-submicrometer analog CMOS technology are discussed in [6,7]. The NBTI induced mismatch in operational amplifier circuits implemented in 90nm technology was found to induce little change in output characteristics compared to time zero process variability for asymmetric stress condition results presented in [8]. Similar circuit level aging simulations under NBTI effect on current mirror, operational amplifier, comparator and digital to analog converter circuits are presented in [9]. The importance of extreme levels of matching for highly accurate applications is highlighted. Effect and modeling of aging induced mismatch in balanced analog circuits is presented in [10–14]. NBTI effect on aging reliability of bandgap reference circuit using thermal sensor DAC is studied in [1] using aging simulation, and the degradation was not found to be significant to cause error to the functionality. NBTI and CHCI aging effects on the performance degradation of RF and analog circuits is presented in [15–19]. An optimum operating voltage that balances NBTI degradation against transistor voltage headroom is presented in [20]. Effect of HCI stress on matching variations is presented in [21]. Analog circuit simulation using the BTI recovery models are shown in [22].

In [5,23] passive techniques using burn-in and calibration to treat aging induced offset in differential structures implemented in 32nm technology are explained. Active NBTI compensation techniques in 65nm technology based on body biasing and differential matching by switching the input pair are presented in [24]. Similar offset voltage reduction of the SRAM sense amplifier implemented in 40nm technology using HCI trimming is proposed in [25]. In [26] the potential to boost the performance of AMS circuits using elevated V_{DD} and thick oxide transistors without degrading the reliability is discussed. A prognostic circuit to measure the threshold voltage shift due to NBTI and report the state of health of analog circuit is reported in [27]. Design tools for efficient analysis and prediction of the lifetime yield of analog circuits together with effects of process variation are presented in [28–32]. A methodology to study the influence of BTI and process variability on functionality of different configurations of an amplifier circuit is investigated in [33].

It can be noticed from the summary above that till date very few publications deal with PBTI effect in nMOSFETs which was introduced with the use of high-κ since 45nm technology node. Moreover AMS circuit reliability evaluation under combined effects of NBTI, PBTI, CHCI and NCHCI is rarely performed [34–36]. For accurate circuit lifetime predictions it is important to consider all the aging effects together since interaction between these effects can enhance or slow down the overall circuit performance degradation depending on the input voltages and stress condition. Also a very limited research is carried out on the countermeasures to compensate and overcome the aging induced performance degradation in the AMS circuits.

1.3 Contributions of This Work

In this investigation non-destructive aging mechanisms associated with the deep-submicrometer CMOS technology and their impact on the end-of-life performance of AMS circuits are evaluated. The combined effects of NBTI, PBTI, CHCI and NCHCI induced transistor aging, resulting mismatch, AMS circuit degradation and contribution of individual aging mechanisms to this performance degradation, for the 32nm high-κ metal gate technology [37] is studied. Instead of studying each aging mechanism independently using dedicated on-chip test structures, the study of its combined effect on circuit performance is proposed since it is important to evaluate how these aging mechanisms interact with one another. Countermeasures to overcome aging induced performance degradation are proposed and demonstrated with measurement results.

An analytical approach to evaluate the contribution of different aging mechanisms to performance degradation of linear circuits is introduced. The importance of circuit topology selection concerning reliability is explained with an example of comparison between two operational amplifier design implementations viz., simple Miller amplifier and folded cascode amplifier. Two solutions viz., chopper stabilization and autozeroing to mitigate the effects of aging induced performance degradation in differential circuits are proposed. The impact of aging degradation on ring oscillator circuit performance is evaluated and an adaptive bipolar tracking technique to monitor and compensate this degradation is demonstrated. Similarly transmission gate switches used commonly in switched capacitor circuits are investigated for degradation due to aging. Based on the aging degradation knowledge of various basic building blocks aging induced performance degradation in complex Nyquist rate and oversampling analog to digital converter circuits are investigated.

1.4 Summary

Although more and more circuit functionality is being shifted into digital domain, analog and mixed signal circuits will always play a very important role in integrated circuits of electronic systems since the real world signals are analog in nature. Hence the importance of analog and mixed signal circuits was highlighted in this chapter. Slowing down of supply voltage and threshold voltage scaling, increase in power consumption, introduction of new gate dielectric stack and continuous scaling of transistor gate length and oxide thickness has enhanced aging mechanisms in deep-submicrometer technology era. The impact of this aging degradation on AMS circuit performance was discussed.

A trend of rising research interest among the research community in the field of AMS circuit reliability was highlighted. And a summary of the current state-of-the-art research in this field was provided. Further introduction to the contributions of this research work to further gain insight into reliability issues in AMS circuits fabricated in 32nm deep-submicrometer CMOS technology were presented. The various topics introduced in this chapter will be covered in more detail in the subsequent chapters.

Chapter 2

Transistor Level Modeling of Aging Mechanisms

At shrinking deep-submicrometer technology nodes severe reliability concerns are raised on device and circuit level. Aging degradation mechanisms such as conducting hot carrier injection (CHCI), non-conducting hot carrier injection (NCHCI) and bias temperature instability (BTI) affect MOSFET parameters like threshold voltage and drain current. These effects are not modeled in conventional transistor models like Spice or Spectre. Thus, the conventional circuit simulators cannot effectively handle aging simulation. Dedicated device models with additional information related to aging degradation mechanisms and reliability simulation tools are therefore required to enable static and dynamic simulations of CHCI, NCHCI and BTI to evaluate device and circuit reliability. This chapter introduces the 32nm high-κ metal gate CMOS technology which was used for aging simulation, test chip implementation and measurements. Further the aging degradation mechanisms treated in this investigation, transistor level modeling of these aging mechanisms and aging simulation flow are explained.

2.1 High-κ Metal Gate CMOS Technology

Starting from 45nm CMOS technology node high-κ (HK) and metal-gate (MG) are used in semiconductor manufacturing process to overcome the problem of increasing leakage current with scaling gate oxide thickness. These HK stacks have less gate leakage because of their higher physical thickness for the same equivalent oxide thickness (EOT), because of the higher dielectric constant (κ) of these materials compared to the conventional SiO_2 or $SiON$ dielectrics. Among various HK materials, Hf-based films of HfO_2 and $HfSiO_2$ are considered as one of the most promising alternative gate dielectrics because of their thermal stability, low density of interface states and low leakage current [37].

However the introduction of HK as new gate dielectric material in combination with MG generates new reliability challenges which were absent in conventional SiO_2 or $SiON$ gate stacks. In addition to negative bias temperature instability (NBTI) in pMOSFET, positive bias temperature instability (PBTI) in nMOSFET is the new reliability concern in HK/MG technology because of the high density of structural defects in the HK dielectrics resulting into additional fast transient charge trapping [38]. Further hot carrier stress has emerged as a dominant degradation factor in short channel MOSFETs [39].

To study the effect of these enhanced reliability issues on analog and mixed signal circuit aging degradation, circuits implemented using state-of-the-art 32nm HK/MG CMOS technology are investigated in this work.

2.2 Aging Degradation Mechanisms

One of the important aspects of any electronic system is to perform reliably for a defined period of time. Various wearout mechanisms can lead to device and circuit degradation over its defined operating lifetime. This results from non-constant aggressive scaling of device dimensions, increasing electric fields and usage of new materials which enhances the reliability concerns at deep-submicrometer CMOS technology era. The aging degradation mechanisms causing wearout of device performances and not hard failure (e.g. dielectric breakdown) are treated in this investigation. In this section these reliability wearout mechanisms are introduced.

2.2.1 Bias Temperature Instability

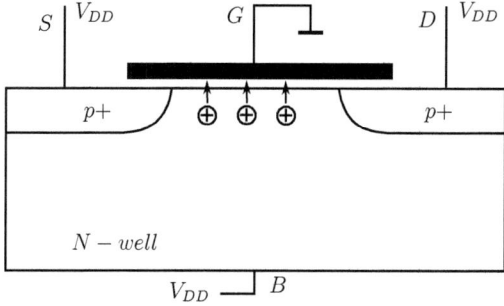

Fig. 2.1: Negative Bias Temperature Instability mechanism

Bias temperature instability is a degradation effect resulting in generation of interface states at the Si/SiO_2 interface and oxide traps under vertical gate oxide fields $F_{ox} \leq 6 - 10 MV/cm$ in inversion ($V_{gs} > |V_{th}|$) and at elevated temperatures (30 to 200°C) [40]. The highest impact of BTI is observed in pMOSFETs when stressed with high negative gate voltage at elevated temperatures [41]. It is referred to as negative BTI (NBTI) due to the negative gate to source voltage. In pMOSFETs, the channel holes interact with the passivated hydrogen bonds in the dielectric resulting into generation on traps and interface states as illustrated in figure 2.1. This results into increase in absolute

Aging Degradation Mechanisms 11

threshold voltage value. The effect of BTI is enhanced at high temperatures. This NBTI aging degradation mechanism effect has been reported 45 years ago [42] and has gained interest due to enhanced reliability concerns in recent years [43–45]. Introduction of new dielectric material like high-κ has enabled BTI effect in nMOSFETs and is referred to as positive bias temperature instability (PBTI) due to positive gate to source voltage. Currently BTI is one of the most serious and important reliability concern for both digital and analog circuits. At advanced technology nodes this effect is enhanced due to reduced voltage headroom, high oxide electric fields resulting from non-constant field scaling, high temperatures due to higher power dissipation and introduction of new dielectric material.

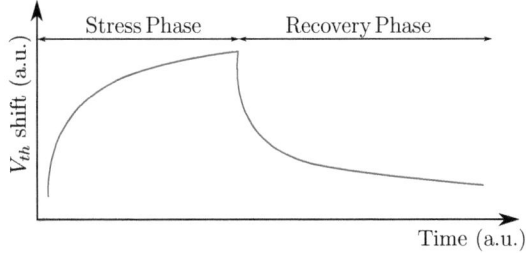

Fig. 2.2: Recovery effect in BTI reliability mechanism [46]

BTI degradation starts relaxing very quickly after the removal of the stress. Figure 2.2 illustrates the recovery of V_{th} drift which is observed once the stress field is removed. This threshold voltage relaxation phenomenons consist of long recovery transients. The recovery component is caused by de-trapping of charge during subsequent removal of stress signal after a stress phase [47]. The resulting recovery consists of fast and slow (so-called permanent) components. These effects have a broad range of time constants, and thus this component can be considered as similar to low frequency ($1/f$) flicker noise [48]. This recovery is beneficial for enhanced device lifetime but it also makes characterization of BTI degradation quite tedious. Recovery after NBTI or PBTI stress in MOSFETs and its dependence of gate voltage, temperature and frequency of stress signal has been a hot topic of research in the past decade [49, 50].

The stress signal causing BTI degradation can be of two types viz., static stress (DC Stress) or dynamic stress (AC Stress). The AC stress is shown to be beneficial for lifetime enhancement of the device and circuit [51–53]. In case of a 50% duty cycle AC stress the recovery occurring during the non-stressing half periods reduces the degradation by around a factor of two compared to the DC stress. Wide research on the topic, dependence of duty cycle, frequency and magnitude of the AC stress on BTI degradation and recovery is available in the literature.

Device degradation due to aging resulting from slow recovering or permanent BTI degradation mainly increases the absolute threshold voltage (V_{th}) of the MOSFETs. This permanent V_{th} shift (($\Delta V_{th})_{N/PBTI}$) behavior in the pMOSFET and nMOSFET transistors resulting from NBTI and PBTI degradation respectively is modeled using equation (2.1), similar to [9, 33, 54] using different set of parameters for pMOSFET and nMOSFET devices. These device related parameters are fitted to single device stress measurements

$$(\Delta V_{th})_{N/PBTI} = A \cdot \left(\frac{V_{gs}}{T_{inv}}\right)^m \cdot e^{\left(\frac{\Delta E}{kT}\right)} \cdot L^\alpha \cdot W^\beta \cdot t^n \quad (2.1)$$

where, A, m, ΔE, α, β are experimentally determined fitting parameters, T_{inv} is the electrically measured oxide thickness, k is the Boltzmann constant, T is the temperature, t is the stress time and n is in the range between 0.19 to 0.26. Different fitting parameters are used to model NBTI in pMOSFET and PBTI in nMOSFET devices.

The ΔV_{th} degradation due to BTI follows a power law behavior in time over wide range of decades. The shift in V_{th} depends on the stress voltage (V_{gs}) with an exponent m and on the stress time (t) with an exponent n. BTI degradation has saturating characteristics at larger stress time. It has a weak dependence to the transistor dimensions (W and L). The temperature (T) dependence is modeled to follow Arrhenius law with the activation energy (ΔE). The BTI degradation also to some extent depends on the drain to source voltage (V_{ds}), however this relation is not modeled here.

The basic BTI mechanism is not yet fully understood. There have been different efforts to model BTI [48, 55, 56]. An accurate but simple to use BTI degradation evaluation and prediction model is required that models not only DC stress degradation and recovery but also the response to dynamic (AC) stress with arbitrary stress/recovery sequences, but at the moment no generally accepted model is available.

2.2.2 Hot Carrier Injection

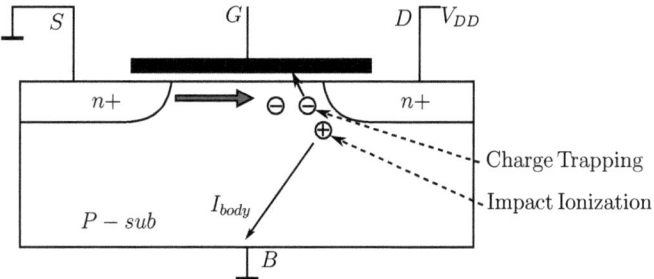

Fig. 2.3: Hot Carrier Injection mechanism

The recent CMOS technologies have witnessed slowing down or stopping of supply voltage (V_{DD}) and threshold voltage (V_{th}) scaling because of the non-scalability of sub-threshold slope whereas the gate length (L) and oxide thickness (t_{ox}) is continuing to scale down. This results in a net increase in lateral electric field, effective channel field and the vertical oxide field. Due to this aggressive non-constant field scaling hot carrier (HC) effect is again a prime concern for device and circuit reliability. The impact of conducting hot carrier injection (CHCI) and non-conducting hot carrier injection (NCHCI) on device

Aging Degradation Mechanisms 13

and circuit reliability are the indicative of the ability of the process to resist wearout mechanisms during operating lifetime [57].

MOSFET devices fabricated in deep-submicrometer CMOS technology era experience very high lateral field and the extension of the pinched-off region approaches the values in the order of the carrier mean free path. Under these conditions carriers can gain large kinetic energies while transiting through the regions of high electric field. When the carrier energy gets significantly larger than that associated with the lattice in thermal equilibrium, they are called hot. These hot carriers can gain enough energy to be injected into the gate oxide, cause interface damage or induce substrate and gate leakage introducing instabilities in the electrical characteristics of a MOSFET device as illustrated in figure 2.3 [40]. These hot carriers in the channel traveling from source to drain can experience very high electric field near the drain region resulting into impact ionization and conducting hot carrier injection (CHCI) into the oxide. Conventional Spice models include impact ionization effect but not CHCI. CHCI leads to degradation of drain current (I_d) and absolute increase in transistor threshold voltage ($|V_{th}|$).

The MOSFETs operating in sub-threshold region or off-state with high drain to source voltage experience interface trap generation in the gate-drain overlap region and localized charge trapping into the spacer oxide [58]. The damage is elevated with increasing temperatures. This effect induces degradation in drain current, I_d and is commonly referred to as off-state or non-conducting HCI (NCHCI).

Device degradation due to aging resulting from CHCI and NCHCI reduces transistor I_d and CHCI also gives a small contribution to the V_{th} shift. The degradation in I_d resulting from CHCI and NCHCI effects are modeled separately by equation (2.2) using different set of parameters for the two effects and also for pMOSFET and nMOSFET devices. These device related parameters are fitted to single device stress measurements

$$(\Delta I_d)_{CHCI/NCHCI} = I_d \cdot B \cdot V_{ds}^p \cdot e^{\left(\frac{\Delta E}{kT}\right)} \cdot L^\delta \cdot t^q \tag{2.2}$$

where, B, p, ΔE, δ are experimentally determined fitting parameters, k is the Boltzmann constant, T is the temperature, t is the stress time and q is in the range between 0.25 to 0.45. Total four sets of different fitting parameters are used to model CHCI and NCHCI degradation in both pMOSFET and nMOSFET devices.

The ΔI_d degradation due to CHCI and NCHCI also follows a power law behavior in time over a wide range of decades. The shift in I_d depends on the stress voltage (V_{ds}) with an exponent p and on the time (t) with an exponent q [59]. This degradation has saturating characteristics at larger stress time. The temperature (T) dependence is modeled to follow Arrhenius law with the activation energy (ΔE). The HCI degradation also to some extent depends on the gate voltage (V_G), however this relation is not modeled here.

To minimize the effect of HCI degradation process modifications like double diffusion of source and drain, and graded drain junctions, etc. are implemented typically [41]. A designer can use larger channel lengths (L) for HCI critical devices.

2.3 Transistor Level Modeling

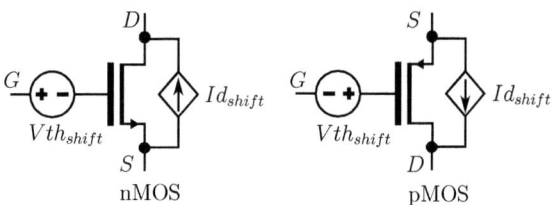

Fig. 2.4: Modeling of MOSFET degradation mechanisms

All MOSFET devices are expected to degrade with time. This causes the circuit performance to deviate from its specifications measured post fabrication. So device and circuit reliability is of prime practical importance in electronic systems. Accurate prediction of aging induced performance degradation is important right from the design phase in order to avoid chip failures at client site and design re-spin. The accuracy of predicting the circuit reliability is closely attached to accurate modeling of aging degradation mechanisms on transistor level. This section explains the modeling of nMOSFET

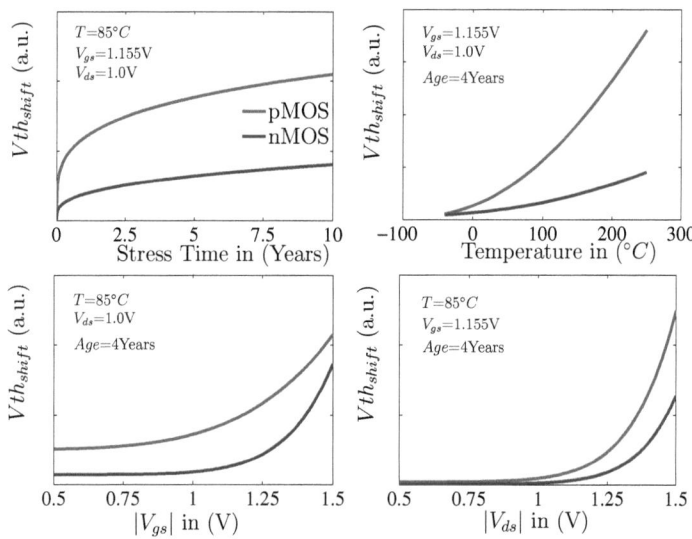

Fig. 2.5: Threshold voltage degradation with respect to stress time, temperature and control voltages simulated using equation (2.3) with different set of fitting parameters for both nMOSFET and pMOSFET devices

Transistor Level Modeling 15

and pMOSFET device aging mechanisms on transistor level in 32nm HK/MG CMOS technology process.

$$Vth_{shift} = a \cdot (\Delta V_{th})_{N/PBTI} + b \cdot (\frac{\Delta I_d}{I_d})_{CHCI} \qquad (2.3)$$

$$Id_{shift} = c \cdot (\frac{\Delta I_d}{I_d})_{CHCI} + d \cdot (\frac{\Delta I_d}{I_d})_{NCHCI} \qquad (2.4)$$

where, a, b, c, d are experimentally determined fitting parameters. Different fitting parameters are used for modeling aging degradation in pMOSFET and nMOSFET devices.

The MOSFET's V_{th} and I_d degradation due to aging mechanisms explained in section 2.2 can together be modeled by replacing the MOSFET model with the equivalent sub-circuit model shown in figure 2.4. The V_{th} degradation is modeled by an equivalent voltage source Vth_{shift} in series to the gate terminal. The I_d degradations due to hot carrier effects are modeled by a current controlled current source (CCCS) between the drain and source terminals, where Id_{shift} is the gain and the current of this source is dependent on the drain current of the MOSFET device. The values of the equivalent sources are determined using equations (2.3)-(2.4), which describe the relation of different contributing factors to the overall degradation. The total Vth_{shift} which is a combination of BTI and CHCI is modeled by equation (2.3). The total Id_{shift} which is a combination of CHCI and NCHCI is modeled by equation (2.4). Equations (2.3)-(2.4) are fitted to

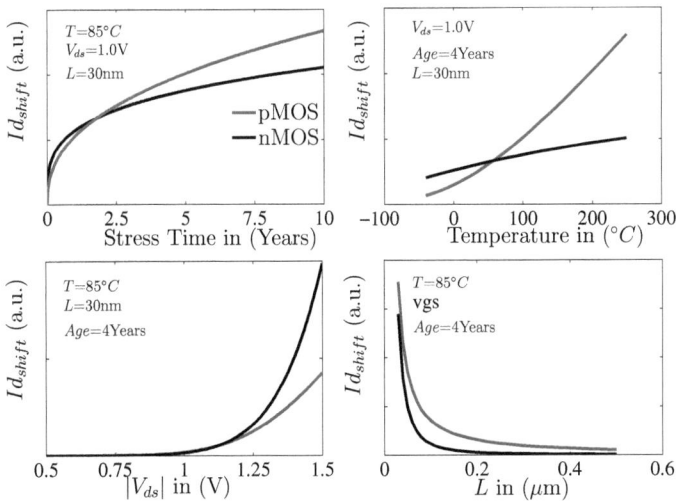

Fig. 2.6: Drain current degradation with respect to stress time, temperature, control voltage and gate length simulated using equation (2.4) with different set of fitting parameters for both nMOSFET and pMOSFET devices

single device stress measurement data using different set of parameters for pMOSFET and nMOSFET devices.

Figure 2.5 plots the relation of Vth_{shift} with respect to stress time (t), temperature (T) and control voltages ($|V_{gs}|$ and $|V_{ds}|$) simulated with equation (2.3) for both nMOSFET and pMOSFET devices with different set of fitting parameters. It can be observed that V_{th} degradation in pMOSFET is high compared to nMOSFET considering combined effect of BTI and CHCI. Figure 2.6 plots the relation of Id_{shift} with respect to stress time (t), temperature (T) and control voltage ($|V_{ds}|$) and transistor gate length (L) simulated with equation (2.4) for both nMOSFET and pMOSFET devices with different set of fitting parameters. A strong dependence of stress conditions is seen in case of degradation induced in pMOSFET and nMOSFET due to CHCI and NCHCI mechanisms.

The transistor level sub-circuit models explained in this section with aging information related to BTI, CHCI and NCHCI stress effects are used in the subsequent chapters for the analytical evaluation of aging degradation, aging simulation and lifetime prediction on circuit level.

2.4 Reliability Simulation Flow

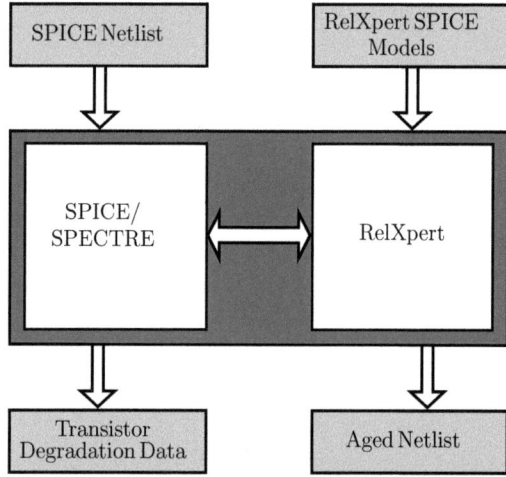

Fig. 2.7: Simulation flow for circuit reliability evaluation

Circuit level aging simulation enables evaluation of the effects of aging degradation on individual transistors of the circuit and the overall circuit performances under DC and AC stress. It indicates the health of the transistors and the circuit after a specified stress time. For these evaluations based on circuit simulation, in this research the aging simulation

Reliability Simulation Flow 17

tool RelXpert^{TM} [60] is used which employs the models explained in section 2.3. In this section an introduction to the general aging simulation flow is provided.

The simulation flow for circuit reliability evaluation under aging degradation is illustrated in figure 2.7. The circuit netlist and the transistor models having additional information related to the parameter values of equations (2.1) and (2.2), describing the aging behavior under BTI and HCI stress are provided to the aging simulation tool. The stress conditions like temperature, stress time (in seconds, minutes or years) and bias conditions are provided in the netlist. The aging simulation tool enables calculation on actual signal waveforms across each transistor in the circuit based on circuit operating conditions. Further these waveforms are used to evaluate the degradation ("Age") in each transistor which can be extrapolated to a specified stress time. In this aging simulation tool the predictions about lifetimes and percentage degradation are made based on formulas presented in section 2.2.

$$Deg(t_i) = \left(A \cdot \left(\frac{V_{gs}}{T_{inv}} \right)^m \cdot e^{\left(\frac{\Delta E}{kT} \right)} \cdot L^\alpha \cdot W^\beta \right)^{\frac{1}{n}} \cdot \left(t_i - t_{(i-1)} \right) \quad (2.5)$$

$$Age = \left(\frac{1}{tsim_{stop} - tsim_{start}} \right) \cdot \int_{tsim_{start}}^{tsim_{stop}} Deg(t_i) dt \quad (2.6)$$

$$\Delta V_{th} = (Age \cdot \mathbf{t})^n \quad (2.7)$$

where Deg is the degradation evaluated during each transient simulation time step (t_i), Age is the degradation of the transistor which can be extrapolated to any stress time ($\mathbf{t}$), $tsim_{start}$ is the start time of the transient simulation, $tsim_{stop}$ is the end time of the transient simulation.

For example to evaluate ΔV_{th} due to BTI degradation for each transistor in the netlist either under DC or AC stress, the aging simulation tool uses Spice/Spectre transient simulations with user specified simulation time via parameters ($tsim_{start}$ and $tsim_{stop}$). To accurately evaluate the circuit aging degradation, simulation for at least the duration of one complete cycle of the input signal must be performed on the circuit under investigation. For each transient simulation time step (t_i) degradation (Deg) in each transistor is evaluated using equation (2.5) which is basically equation (2.1) without stress time ($\mathbf{t}$) dependence multiplied with the size of each time step. Finally at the end of this transient simulation, equations (2.6) evaluates the value of degradation (Age) for each transistor, which can be used to evaluate ΔV_{th} at any specified stress time $\mathbf{t}$ using equation (2.7). Similarly degradation due to other stress mechanisms is evaluated. To handle the recovery effect in case of AC stress, an AC factor (< 1) is multiplied to the BTI degradation results depending on the duty cycle of the stress signal waveforms across each transistor.

After evaluating the BTI and HCI degradation for each individual MOSFET device in the circuit, the aged devices in the circuit netlist are replaced with the sub-circuit model described in section 2.3. On this aged circuit netlist transient simulation can then be performed for comparison of before and after degradation waveforms. It provides information on how long the circuit takes to degrade to a certain degradation percentage and the degradation of the chosen circuit at the selected circuit age. Further the aging simulation tool lists for all measured transistors the information about their "Age" degradation. This transistor degradation data summary enables tracking the device with maximum HCI or BTI effect. It helps to detect and identify weakest spots in the circuit and work on them to improve the circuit reliability.

2.5 Summary

Accurate transistor aging models and aging simulation tools are needed to correctly predict and analyze the circuit performance degradation over lifetime or at end-of-life conditions. In this chapter the reliability issues related to the state-of-the-art high-κ metal gate CMOS technology were discussed. The reliability wearout mechanisms treated in this work viz., BTI and HCI were introduced. Aging leads to shift in threshold voltage and drain current of MOSFET devices and this degradation depends mainly on applied bias conditions, temperature and stress time. BTI mechanism has weak dependence of device dimensions whereas HCI degradation can be reduced by increasing the transistor channel length. The modeling of the degradation induced by these aging mechanisms on transistor level based on single device measurements was explained using sub-circuit models for nMOSFET and pMOSFET devices. And an aging simulation flow to evaluate the reliability of the circuit post aging was discussed.

Chapter 3
Circuit Level Analytical Evaluation and Accelerated Aging

Electronic systems must meet expected performance specifications over a desired product lifetime. Evaluation of the impact of aging on circuits is not trivial due to superposition of different aging mechanisms which can enhance or slow down the aging induced performance degradation. In chapter 2 the MOSFET device level models with aging induced parameter drifts and reliability simulation flow are discussed. Degradation due to different reliability mechanisms like positive bias temperature instability (PBTI), negative bias temperature instability (NBTI), conducting hot carrier injection (CHCI) and non-conducting hot carrier injection (NCHCI) are taken into consideration. Based on these models and aging simulation tools, in this chapter an approach to analytically evaluate aging degradation in linear analog and mixed signal circuits is introduced [61]. The method introduced in the chapter is based on using sensitivity analysis. It is fast, intuitive and gives quantitative insight into the various factors contributing to circuit aging.

Another important aspect of reliability study of an electronic system over its lifetime is to find a meaningful stress condition to shorten the product life span from 4 years end-of-life (EoL) use case to 2 days or less. This enables circuit reliability experiments to be performed in a reasonable time and allows quick feedback. This concept of shrinking the use case product EoL for reliability studies is known as accelerated aging [40]. Again finding a right accelerated EoL stress condition to map with the use case life span of the circuit is not trivial since its necessary to ensure that the contribution of reliability mechanisms match in both the cases. Also it is necessary to ensure that no new reliability mechanism is introduced after shrinking the product lifetime. The concept of evaluating such accelerated stress condition [34] is summarized in this chapter since it is used while performing measurements in later chapters. Also the methodology for analytical evaluation introduced in this chapter is further used for mapping of realistic EoL aging behavior for an accelerated stress test setup.

3.1 Methodology for Analytical Evaluation

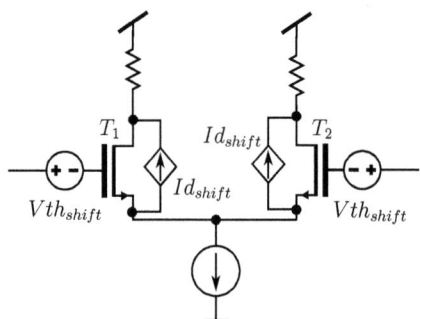

Fig. 3.1: Aging degradation induced parameter drifts in MOSFET devices

In analog and mixed signal (AMS) circuits aging degradation mechanisms significantly affect MOSFET device parameters like threshold voltage (V_{th}) and drain current (I_d) as illustrated in figure 3.1, resulting into degradation of circuit performance over product lifetime. It was explained in section 2.2 that the contribution of different aging mechanisms to these device parameter drifts depends on the applied stress condition. Due to interaction of these aging mechanisms, the impact of aging induced device parameter drifts either slows down or enhances circuit performance degradation. It is important to note that the transistors with maximum parameter drifts are not always the weakest spots in circuit design since degradation of circuit performance depends on the sensitivity of the performance towards that parameter drift. Further while evaluating circuit performance degradation in differential structures, aging induced mismatch in matched pairs is more important compared to individual transistor parameter drift. So to study the behavior, sensitivity and contribution of different aging mechanisms on circuit performance an analytical approach is required. The methodology to analytically evaluate circuit level aging induced performance degradation is explained next.

3.1.1 Steps for Analytical Evaluation

The methodology to analytically evaluate performance degradation due to aging mechanisms in linear AMS circuits can be divided into following three simple steps:

1. The sensitivity of the circuit performance under investigation, towards each transistor's V_{th} and I_d shift is evaluated. This can be done using hand calculations based on small signal analysis approach. However it gets tedious and time consuming for complex circuits with large number of transistors. Another approach which makes this evaluation trivial is one using the sensitivity analysis simulation option which is available in most of the standard Spice simulation tools. This option finds the

Methodology for Analytical Evaluation 21

sensitivities of the defined output variable or performance with respect to the component instance parameters in the circuit. The typical command for the Spectre simulation tool [62] to perform sensitivity analysis is

sens (*output variable or performance*) to (*design parameters*) for (*analysis*)
E.g. sens ($OFFSET$) for (DC)

For this example the DC sensitivities of circuit performance OFFSET are evaluated with respect to selected component instance model parameters of all devices in the design which also includes aging induced V_{th} and I_d shift.

2. The reliability simulation flow introduced in section 2.4 is used to evaluate aging degradation in individual transistors of the circuit depending on the applied stress condition. Aging simulation tools such as RelXpert [60] generate a report which contains information regarding NBTI, PBTI, CHCI and NCHCI degradation in each transistor for the simulated stress condition. These evaluations are based on the equations (2.1)-(2.2) discussed in section 2.2. Using this aging information and equations (2.3)- (2.4) contribution of NBTI, PBTI and CHCI to V_{th} shift and contribution of CHCI and NCHCI to I_d shift is evaluated for each transistor of the circuit.

3. The results obtained from steps 1 and 2 above i.e. sensitivities of investigated circuit performance towards all device parameter shifts and the contribution of different aging mechanisms to device parameter shifts for each transistor, are multiplied together. Due to linearity of the circuit, when these multiplication results evaluated for all transistors are added up, it determines the investigated circuit performance degradation.

This analysis provides insight into contribution of various aging reliability mechanisms toward circuit performance degradation and helps to identify the weak spots in the design concerning reliability.

3.1.2 Application of Methodology

In this section the methodology explained in section 3.1.1 is applied in practice to an example of an operational transconductance amplifier (OTA) circuit configured in a open loop configuration. It is basically a two stage fully differential amplifier with frequency compensation. The aim of this example study is to demonstrate the effectiveness of the introduced methodology in determining the contribution of different aging mechanisms to aging induced performance degradation of an OTA circuit. The schematic of the designed OTA is illustrated in figure 3.2. The OTA circuit is designed in a 32nm high-κ, metal gate CMOS technology.

This aging simulation is performed for a mobile phone EoL use case condition of 4 years, 85°C and 105% of worst case V_{DD} ($V_{DD} = 1.155V$), under asymmetrical DC input stress ($Vin_n = 1.155V$, $Vin_p = 0V$). Moreover all the reliability mechanisms discussed before i.e. NBTI, PBTI, CHCI and NCHCI are taken into account. In this case the worst affected circuit performance due to aging induced parameter drifts is found

Chapter 3. Circuit Level Analytical Evaluation and Accelerated Aging

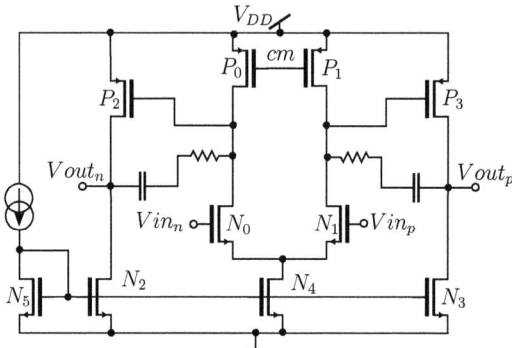

Fig. 3.2: Schematic of simple Miller OTA circuit

to be offset whereas other performances like amplifier gain, bandwidth and phase margin are not significantly affected. The table 3.1 summarizes the simulated performances of the simple Miller OTA circuit before and after stress evaluated at $V_{DD} = 1V$ and $T = 25°C$. To study the contribution of each transistor's parameter shift resulting from different aging mechanisms to offset degradation in the OTA circuit, the procedure explained in section 3.1.1 is followed.

$$V_{os} = Vout_p - Vout_n; @Vin_n = Vin_p = V_{cm} \qquad (3.1)$$

$$V_{is} = \frac{V_{os}}{A_o} \qquad (3.2)$$

where V_{os} is the output referred offset, V_{cm} is the common mode voltage (analog ground), V_{is} is the input referred offset and A_o is the open loop amplifier DC gain

In the OTA example, output referred offset (V_{os}) given by equation (3.1), resulting from asymmetrical stress is considered to be the dominant aging induced performance degradation and hence its sensitivity towards aging induced V_{th} and I_d shift of each transistor is evaluated. The sensitivity analysis option available in standard simulation tools is used to evaluate the sensitivity of V_{os} towards V_{th} and I_d shift of each transistor. The results are illustrated in figure 3.3 and figure 3.4 respectively.

Next, the degradations of individual transistors in the circuit due to aging mecha-

Stress Condition	Gain (dB)	Phase margin (°)	Bandwidth (MHz)	Input referred offset (mV)
Virgin (Before Stress)	53.43	88.56	741.4	0
4Yrs, 85°C, $V_{DD} = 1.155V$	53.41	89	741.35	5.29

Table 3.1: Simulated performances of the simple Miller OTA circuit evaluated at $V_{DD} = 1V$ and $T = 25°C$

Methodology for Analytical Evaluation 23

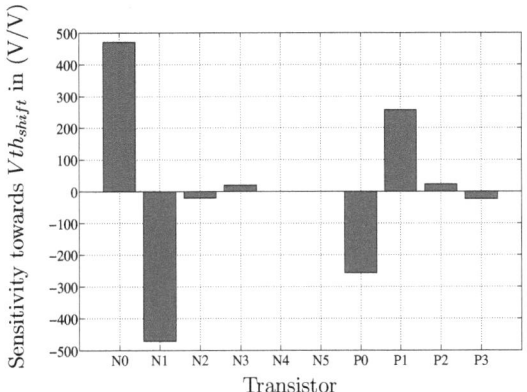

Fig. 3.3: Sensitivity of output referred offset towards aging induced Vth_{shift} of different transistors in the simple Miller OTA circuit

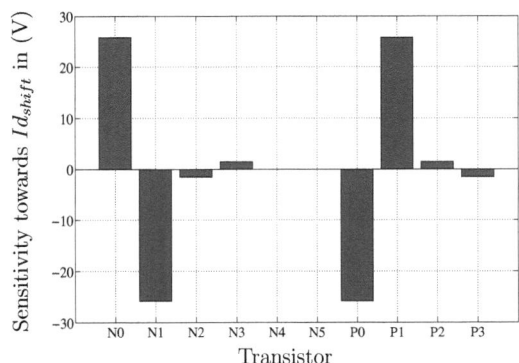

Fig. 3.4: Sensitivity of output referred offset towards aging induced Id_{shift} of different transistors in the simple Miller OTA circuit

nisms like PBTI, NBTI, CHCI and NCHCI are noted from the output generated by the aging simulation tool. Using (2.3) and (2.4) the contributions of different aging mechanisms to V_{th} and I_d shift of each transistor are evaluated with different parameters for nMOSFET and pMOSFET devices. The results are illustrated in figure 3.5 and figure 3.6 respectively. The asymmetrical stress condition causes transistors N_0 and P_2 to experience perfect BTI stress condition: high gate source voltage and low drain source voltage, and transistor P_0 to experience perfect CHCI stress condition: high drain source voltage.

24   Chapter 3. Circuit Level Analytical Evaluation and Accelerated Aging

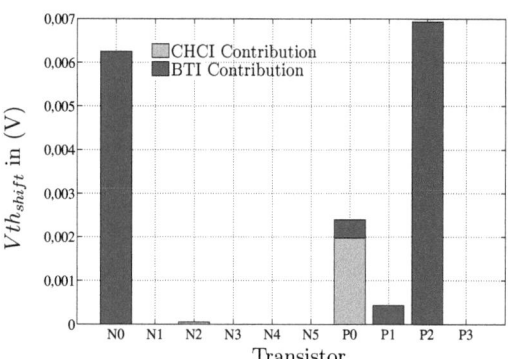

Fig. 3.5: Transistor level Vth_{shift} resulting in open loop simple Miller OTA circuit after aging simulation with stress time of 4 Yrs at $V_{DD} = 1.155V$ and $T = 85°C$

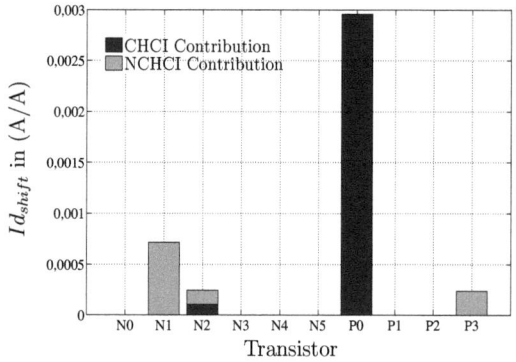

Fig. 3.6: Transistor level Id_{shift} resulting in open loop simple Miller OTA circuit after aging simulation with stress time of 4 Yrs at $V_{DD} = 1.155V$ and $T = 85°C$

$$V_{os} = S_{V_{T1}} \cdot (a \cdot (\Delta V_{th})_{N/PBTI} + b \cdot (\frac{\Delta I_d}{I_d})_{CHCI})_{T1} + \ldots$$
$$+ S_{I_{T1}} \cdot (c \cdot (\frac{\Delta I_d}{I_d})_{CHCI} + d \cdot (\frac{\Delta I_d}{I_d})_{NCHCI})_{T1} + \ldots \quad (3.3)$$
$$= S_{V_{T1}} \cdot Vth_{shift_{T1}} + S_{V_{T2}} \cdot Vth_{shift_{T2}} + \ldots$$
$$+ S_{I_{T1}} \cdot Id_{shift_{T1}} + S_{I_{T2}} \cdot Id_{shift_{T2}} + \ldots \quad (3.4)$$

The V_{os} can then be evaluated analytically using equation (3.4) which is obtained from equation (3.3). For each transistor (T_n), the sensitivity $(S_{V_{Tn}})$ of V_{os} towards V_{th} shift

Methodology for Analytical Evaluation

as well the sensitivity $(S_{I_{T_n}})$ towards I_d shift is multiplied with the respective V_{th} and I_d shifts. These results are illustrated in figure 3.7 and figure 3.8 respectively. Finally these multiplication results representing individual transistor contributions are summed up to get the overall aging induced offset. The results of this analysis are matching closely with results obtained by circuit simulation using RelXpert, as illustrated in table 3.2. Thus the contributions of individual aging mechanisms to overall circuit performance degradation are analyzed and evaluated.

Stress Condition	Using Simulation	Using Analytical Method
4Yrs, 85°C, $V_{DD} = 1.155V$	5.29	5.32

Table 3.2: **Comparison between simulated and analytically evaluated aging induced input referred offset in** (mV) **resulting in the open loop simple Miller OTA circuit evaluated at** $V_{DD} = 1V$ **and** $T = 25°C$

3.1.3 Insight into Aging Degradation Mechanisms

The insight gained from the analytical evaluation of aging degradation mechanisms involved in performance degradation of an OTA circuit under asymmetrical stress is discussed in this section. It can be observed from figure 3.3 that aging induced output referred offset is most sensitive to mismatch in V_{th} shift of the input differential pair transistors (N_0 and N_1) in the OTA circuit. Analysis of data from aging simulation in figure 3.5 and figure 3.6 illustrates that V_{th} shift is a combined effect of BTI and CHCI degradation which is evaluated using (2.3), and the I_d shift is a combined effect of CHCI and NCHCI which is evaluated using (2.4). Further output transistor P_2 experiences most

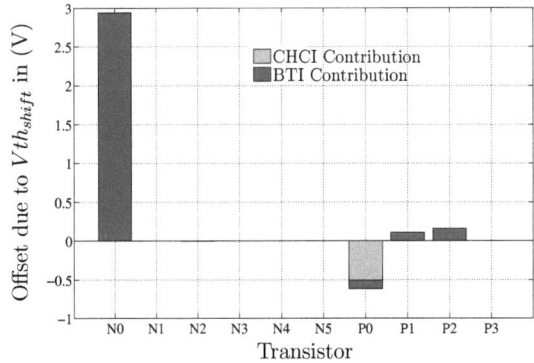

Fig. 3.7: Output referred offset resulting in open loop simple Miller OTA circuit due to transistor level Vth_{shift} after aging simulation with stress time of 4Yrs at $V_{DD} = 1.155V$ and $T = 85°C$

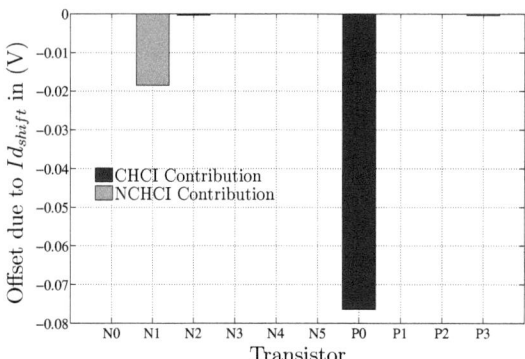

Fig. 3.8: Output referred offset resulting in open loop simple Miller OTA circuit due to transistor level Id_{shift} after aging simulation with stress time of 4Yrs at $V_{DD} = 1.155V$ and $T = 85°C$

V_{th} shift due to high NBTI stress. However, figure 3.7 shows that the dominant contribution to V_{os} due to V_{th} shift comes from input transistor N_0 due to high sensitivity of V_{os} to large mismatch in input transistor pair resulting from PBTI stress induced high V_{th} shift in N_0 and NCHCI stress induced small I_d shift in N_1. NBTI degradation of pMOSFET devices P_0 and P_1 is equal since they see same gate to source voltage from common mode (cm) feedback circuit (neglecting the V_{ds} dependence) and hence compensates each other due to differential signaling. Similarly, figure 3.8 shows that the dominant contribution to V_{os} due to I_d shift comes from CHCI degradation of pMOSFET device P_0. Further, it was observed that V_{th} shift makes the dominant contribution (96%) to V_{os} compared to I_d shift (4%).

3.2 Concept of Accelerated Aging

In order to perform quick measurements to evaluate the lifetime reliability of an integrated circuit, it is necessary to map the use case EoL stress condition of the product to an accelerated stress condition. From equations (2.1) and (2.2) it can be noted that the shrinking in lifetime of a MOSFET device is possible by increasing the temperature (T) and/or the bias voltages; gate to source voltage ($|V_{gs}|$) for BTI and drain to source voltage ($|V_{ds}|$) for HCI. However while shrinking lifetime on a circuit level care must be taken that this mapping does not introduce new reliability mechanisms and the contribution of existing reliability mechanisms remain comparable in both cases. In [34] the concept of deriving an accelerated test setup from the dominating degradation effect is introduced. It is shown that a general increase in temperature and voltage leads to significant deviation in the lifetime degradation of a circuit compared to that under use case conditions. However the evaluation of an accelerated stress condition based on a dominant reliability

Concept of Accelerated Aging

mechanism greatly enhances the mapping of lifetime degradation.

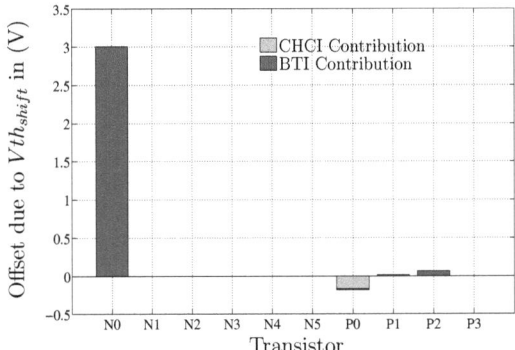

Fig. 3.9: Output referred offset resulting in open loop simple Miller OTA circuit due to transistor level Vth_{shift} after aging simulation with stress time of 10^3s at $V_{DD} = 1.394V$ and $T = 125°C$

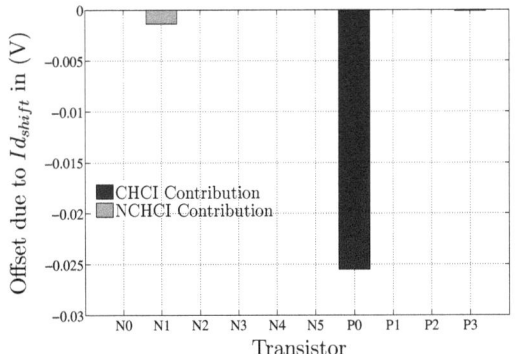

Fig. 3.10: Output referred offset resulting in open loop simple Miller OTA circuit due to transistor level Id_{shift} after aging simulation with stress time of 10^3s at $V_{DD} = 1.394V$ and $T = 125°C$

Using the analytical evaluation methodology discussed in section 3.1 and an accelerated test mapping concept from [34], a realistic mapping of mobile phone EoL use case conditions (4 Years, 85°C, 105% of worst case V_{DD}) for the introduced OTA circuit onto an accelerated measurement test setup of $10^3 s$ can be determined. The procedure is explained in the next paragraph.

Stress Time	T (°C)	V_{scale} (%)	Offset (mV)
4Yrs	85	105% (V_{DD}=1.155V)	5.29
$10^3 s$	125	126.73% (V_{DD}=1.394V)	6.099

Table 3.3: Comparison between simulated aging induced input referred offset resulting from EoL use case condition with respect to mapped acceleration condition in the open loop simple Miller OTA configuration evaluated at $V_{DD} = 1V$ and $T = 25°C$

Based on the analysis presented in section 3.1.3 it is clear that PBTI in nMOSFET device N_0 is the dominant reliability mechanism contributing to aging induced offset in the OTA circuit. Hence for stress time of $10^3 s$ and stress temperature of 125°C, a voltage scaling factor (V_{scale}) was evaluated using equations (2.1) such that equation (3.5) is satisfied, in order to find a realistic acceleration test setup. The results of this mapping are summarized in table 3.3.

$$Vth_{shift}(V_{DDusecase}, T_{usecase}, 4Yrs) = Vth_{shift}(V_{DDaccelerated}, T_{accelerated}, 10^3 s) \quad (3.5)$$

Figure 3.9 and 3.10 illustrates the contribution of various reliability mechanisms towards aging induced output referred offset evaluated for accelerated stress condition of $10^3 s$, 125°C at 126.73% of worst case V_{DD}. Comparing figure 3.9 with figure 3.7 and figure 3.10 with figure 3.8, it is seen that the order of relevance of degradation mechanisms remains the same under acceleration. The deviation between resulting offset is small (13.26%) and hence this voltage and temperature scaling can be used to evaluate degradation of the OTA circuit equivalent to 4 Years aging at $V_{DD} = 1.155V$ (105% of worst case V_{DD}) and $T = 85°C$.

3.3 Summary

In this chapter a new methodology to analytically evaluate aging degradation of linear analog and mixed signal circuits was proposed and discussed. It was based on a three step approach: evaluating the sensitivity of a circuit performance towards degrading device parameters by sensitivity analysis, evaluating the contributions of different degradation mechanisms and summing up the product of performance sensitivity with respective contributions due to different degradation mechanisms. It proved to be in good agreement with circuit simulations, but with considerably less computing effort and providing more intuitive insight into the various degradation contributions. This method was demonstrated for an amplifier circuit designed in a 32nm high-κ, metal gate CMOS technology. This circuit was analyzed for reliability after aging by simulation and using the proposed methodology. It was observed that the most severely affected performance due to aging was amplifier offset, whereas other performances like amplifier gain, bandwidth and phase margin are not affected. It was highlighted that for differential circuits aging induced mismatch in matched pairs degrades the circuit performance. Hence asymmetrical stress is most harmful for reliability of such circuits. The transistor most affected by aging

Summary

degradation is not always the weakest spot concerning reliability in circuits. The circuit performance is most affected by the aging induced mismatch in the matched transistor pairs toward which the performance under investigation has highest sensitivity.

Further, the concept of accelerated aging to perform quick circuit lifetime prediction measurements under aging degradation was introduced. With the results of the sensitivity analysis a realistic mapping of the circuit aging onto an accelerated measurement test setup was determined.

Chapter 4
Aging in Operational Amplifiers

4.1 Operational Amplifiers

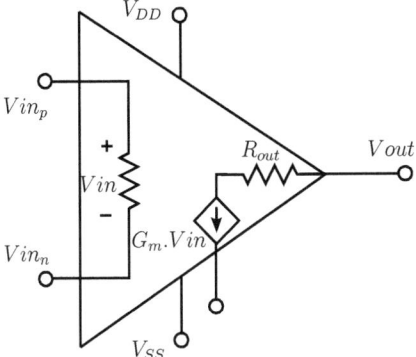

Fig. 4.1: Equivalent circuit for an operational amplifier

An operational transconductance amplifier (OTA) is one of the most important building blocks in analog and mixed signal circuits. The equivalent circuit for an OTA is illustrated in figure 4.1. A high gain OTA is typically used in negative feedback configuration to achieve a precise closed loop transfer function ideally independent of the OTA's open loop gain. It is also used in open loop configuration without frequency compensation as pre-amplifier stage in the latch based comparator circuits. Typically fully differential OTA topologies are used because of its advantages like high DC and dynamic common mode rejection, increased output voltage swing, increased immunity to external noise and reduced even order harmonics.

In this chapter first the findings related to aging in closed and open loop OTA circuit configurations [3, 34, 61, 63] are summarized. Based on the methodology explained in chapter 3, contribution of different aging mechanisms towards aging induced performance degradation in these two configurations are discussed. Next, two OTA design implementations viz., simple Miller OTA and folded cascode OTA are compared for aging induced performance degradation. The advantages of cascode structures towards shielding of transistors from high bias voltages are highlighted and on the other hand the difficulties in design of such topologies at reduced supply voltage are discussed. All the OTA circuits are designed in a 32nm high-κ, metal gate CMOS technology and the investigations are carried out under asymmetrical input stress condition since it induces maximum degradation in matched differential structures.

4.1.1 Closed Loop OTA

Figure 4.2 illustrates the schematic of closed loop OTA circuit configuration in fully differential topology. This is the most commonly used configuration for OTA circuits

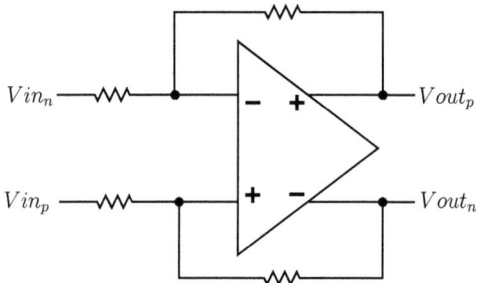

Fig. 4.2: Schematic of OTA circuit in closed loop configuration

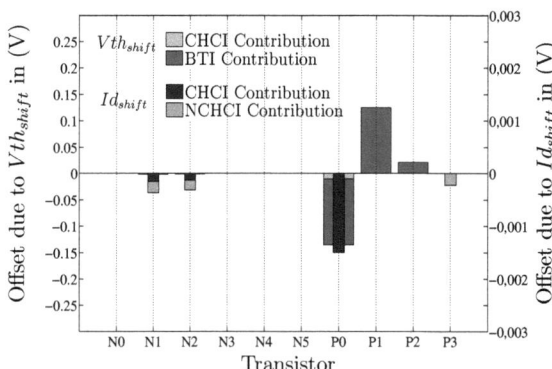

Fig. 4.3: Output referred offset resulting in closed loop simple Miller OTA circuit due to transistor level Vth_{shift} and Id_{shift} after aging simulation with stress time of 4 Yrs at $V_{DD} = 1.155V$ and $T = 85°C$

Operational Amplifiers 33

Stress Time	T (°C)	V_{scale} (%)	Offset (mV)
4 Yrs	85	105% (V_{DD} = 1.155V)	0.012
10 Yrs	125	105% (V_{DD} = 1.155V)	0.0305

Table 4.1: Simulated aging induced input referred offset resulting from asymmetrical DC input stress ($Vin_n = 1.155V$, $Vin_p = 0V$) in closed loop simple Miller OTA configuration evaluated at $V_{DD} = 1V$ and $T = 25°C$

either as unity gain buffers or amplifiers with precise gain. Considering the OTA circuit in figure 3.2, under closed loop condition the input transistor pair operate near common mode voltage and hence do not experience high stress voltage. However the output transistors can see full supply voltage levels. Since the sensitivity of the offset is small for the parameter drifts in output transistor pair as illustrated in figures 3.3 and 3.4, the aging induced offset is small even for worst case asymmetrical input stress conditions. Aging induced input referred offset in closed loop OTA configuration for a mobile phone EoL use case condition of 4 years, 85°C and 105% of worst case V_{DD} ($V_{DD} = 1.155V$), under asymmetrical DC input stress ($Vin_n = 1.155V$, $Vin_p = 0V$) is presented in table 4.1. Figure 4.3 confirms that the dominant contribution to aging induced output referred offset due to V_{th} shift comes from NBTI stress mechanism in output pMOSFET device P_2. NBTI degradation of pMOSFET devices P_0 and P_1 compensates each other since they see same gate to source voltage from common mode (cm) feedback circuit.

4.1.2 Open Loop OTA

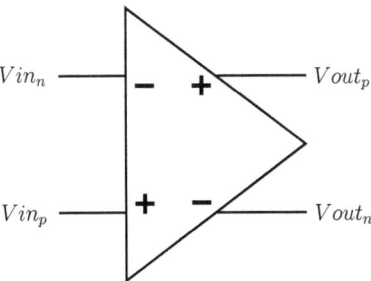

Fig. 4.4: Schematic of OTA circuit in open loop configuration

Figure 4.4 illustrates the schematic of an open loop OTA circuit configuration in fully differential topology. Again considering the same OTA circuit in figure 3.2, under open loop condition both the input and output transistor pairs of the OTA circuit can see full supply voltage levels. The sensitivity of the offset is very high for the mismatch in parameter drifts in input transistor pair as illustrated in figure 3.3. Hence worst case offset results due to high asymmetrical input stress conditions. Aging induced input referred offset in open loop OTA configuration for a mobile phone EoL use case condition

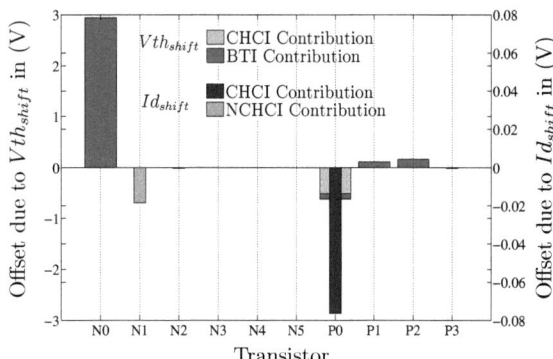

Fig. 4.5: Output referred offset resulting in open loop simple miller OTA circuit due to transistor level Vth_{shift} and Id_{shift} after aging simulation with stress time of 4 Yrs at $V_{DD} = 1.155V$ and $T = 85°C$

of 4 years, 85°C and 105% of worst case V_{DD} ($V_{DD} = 1.155V$), under asymmetrical DC input stress ($Vin_n = 1.155V$, $Vin_p = 0V$), is presented in table 4.2. Here, the input referred offset is evaluated using equation (3.1). Figure 4.5 confirms that the dominant contribution to induced output referred offset due to V_{th} shift comes from PBTI reliability mechanism in input nMOSFET device N_0.

Comparing the simulated results in table 4.1 and table 4.2, it can be noted that under asymmetrical stress conditions open loop OTA configurations are significantly affected by aging in terms of induced offset compared to closed loop configuration. In the OTA circuit (figure 3.2) the bias current through transistors N_4 and N_5 decides the gain, phase margin and bandwidth. Both in closed and open loop circuit configurations these transistors see only limited gate and drain voltages, so no remarkable degradation due to aging mechanisms occurs. Consequently, no significant performance degradation of gain, phase margin and bandwidth (less than 1%) is observed after aging since the bias current is not affected. This is true however only when all the transistor operate in saturation region even after parameter drifts due to aging. This behavior also shows that the current mirror structures are robust towards aging effects.

Stress Time	T (°C)	V_{scale} (%)	Offset (mV)
4 Yrs	85	105% ($V_{DD} = 1.155V$)	5.29
10 Yrs	125	105% ($V_{DD} = 1.155V$)	10.438

Table 4.2: Simulated aging induced input referred offset resulting from asymmetrical DC input stress ($Vin_n = 1.155V$, $Vin_p = 0V$) in open loop simple Miller OTA configuration evaluated at $V_{DD} = 1V$ and $T = 25°C$

4.2 Comparison Between Aging of Different OTA Topologies

In order to study the behavior of circuit level performance degradation due to aging mechanisms on different OTA designs, in this section a comparison between two OTA circuit implementations is presented. Two very popular fully differential OTA topologies viz., simple two stage Miller OTA and folded cascode two stage OTA circuits are studied. The details of their implementation and performance degradation due to aging under mobile phone EoL use case condition are presented in this section.

4.2.1 Simple Miller OTA

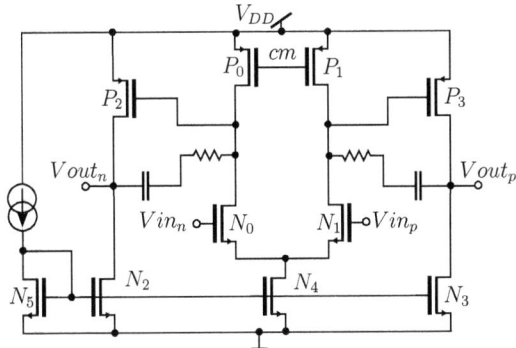

Fig. 4.6: Schematic of simple Miller OTA circuit

DC Gain = 53.43 dB	Gain Bandwidth = 741.4 MHz
Phase Margin = 88.56°	Offset = 0V

Table 4.3: Simulated simple Miller OTA performances evaluated at V_{DD}=1V and T=25°C

The schematic of two stage simple Miller OTA in fully differential configuration and with frequency compensation is illustrated in figure 4.6. It is basically the same circuit discussed in section 3.1.2. Simulated performances of the simple Miller OTA at nominal condition (Virgin, $T = 25°C$ and $V_{DD} = 1V$) are listed in table 4.3. Implementing this circuit using only regular threshold voltage (V_{th}) pMOSFET and nMOSFET devices with $V_{th} \approx 0.45V$ and supply $V_{DD} = 1V$ leads to significant challenges in using input signals with the common mode voltage $V_{cm} = 0.5V$ at the gate of input differential pair, N_0 and N_1, and maintaining the tail transistor N_4 in saturation. To overcome this problem transistors N_0 and N_1 are implemented using low V_{th} nMOSFET devices and transistor N_4 using regular V_{th} nMOSFET device sized to achieve a low overdrive voltage.

Sensitivity of output referred offset towards aging induced Vth_{shift} and Id_{shift}, and the contribution of various aging mechanisms towards aging induced offset is presented in section 3.1.2. Aging induced input referred offset in open loop simple miller OTA configuration for a mobile phone EoL use case condition of 4 years, 85°C and 105% of worst case V_{DD} ($V_{DD} = 1.155V$), under asymmetrical DC input stress ($Vin_n = 1.155V$, $Vin_p = 0V$) is presented in table 4.2. And the insight gained from the analytical evaluation of aging degradation mechanisms involved in performance degradation of this simple miller OTA circuit under asymmetrical stress is discussed in section 3.1.3.

4.2.2 Folded Cascode OTA

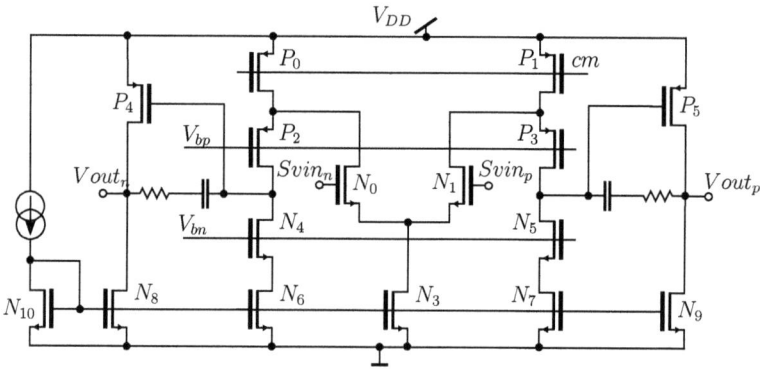

Fig. 4.7: Schematic of folded cascode OTA circuit

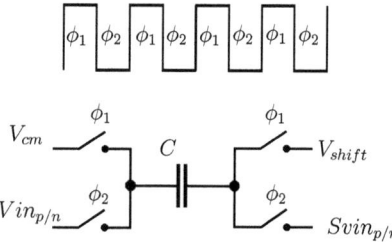

Fig. 4.8: DC level shifting circuit used to shift the V_{cm} of the input signals

The schematic of two stage folded cascode OTA with frequency compensation in fully differential configuration is illustrated in figure 4.7. Simulated performances of the folded cascode OTA circuit at nominal condition (before stress, $T = 25°C$ and $V_{DD} = 1V$) are listed in table 4.4. This circuit is implemented using only regular threshold voltage pMOSFET and nMOSFET devices with $V_{th} \approx 0.45V$ and supply $V_{DD} = 1V$. The

Comparison Between Aging of Different OTA Topologies 37

DC Gain = 59.71 dB	Gain Bandwidth = 736.8 MHz
Phase Margin = 89°	$V_{offset} = 0V$

Table 4.4: Simulated folded cascode OTA performances evaluated at $V_{DD} = 1V$ and $T = 25°C$

Stress Time	T (°C)	V_{scale} (%)	Offset (mV)
4 Yrs	85	105% ($V_{DD} = 1.155V$)	0.775
10 Yrs	125	105% ($V_{DD} = 1.155V$)	1.39

Table 4.5: Simulated aging induced input referred offset resulting from asymmetrical DC input stress ($Vin_n = 1.155V$, $Vin_p = 0V$) in open loop folded cascode OTA configuration evaluated at $V_{DD} = 1V$ and $T = 25°C$

challenge in using the input signals with the common mode voltage $V_{cm} = \frac{V_{DD}}{2}$ at the gate of input differential pair, N_0 and N_1, and maintaining the tail transistor N_3 in saturation is overcome with a simple level shifting circuit as depicted in figure 4.8. This circuit is added before the inputs of this OTA topology. This simple level shifting circuit shifts each input signal ($Vin_{p/n}$) of the OTA with $V_{cm} = \frac{V_{DD}}{2}$ to a new shifted signal ($Svin p/n$) with higher V_{cm} depending on the difference between the common mode and the shift voltage (V_{shift}). During clock phase when ϕ_1 is ON and ϕ_2 is OFF, the difference between V_{shift} and V_{cm} is stored onto the capacitor. And during the phase when ϕ_1 is OFF and ϕ_2 is ON, this difference adds with $Vin_{p/n}$ to get a DC shifted signal $Svin p/n$.

The aging simulations on this folded cascode OTA topology in open loop configuration are performed for the same mobile phone EoL use case of 4 years, 85°C and 105% of worst case V_{DD} ($V_{DD} = 1.155V$), under asymmetrical DC input stress ($Vin_n = 1.155V$, $Vin_p = 0V$). In this case the worst affected circuit performance due to aging induced parameter drift is also found to be offset. Table 4.5 summarizes the simulated input referred offset after stress evaluated at $V_{DD} = 1V$ and $T = 25°C$. An analytical evaluation and insight into aging mechanisms in the folded cascode OTA is explained next based on the methodology introduced in chapter 3.

The sensitivity of output referred offset (V_{os}) towards V_{th} and I_d shift of each transistor is evaluated. The results are illustrated in figure 4.9 and figure 4.10 respectively. Next, the degradation of individual transistors in the circuit due to aging mechanisms like PBTI, NBTI, CHCI and NCHCI are noted from the output generated by the aging simulation tool. And using (2.3) and (2.4) the contribution of different aging mechanisms to V_{th} and I_d shift of each transistor is evaluated with different parameters for nMOSFET and pMOSFET devices. The results are illustrated in figure 4.11 and figure 4.12 respectively. The V_{os} is then evaluated analytically using equation (3.4). The contribution of different aging mechanisms causing V_{th} and I_d shifts in individual transistors which results into induced offset are illustrated in figure 4.13 and figure 4.14 respectively. The results of this analysis are matching closely with results obtained by circuit simulation using RelXpert, as illustrated in table 4.6. Thus the contributions of individual aging

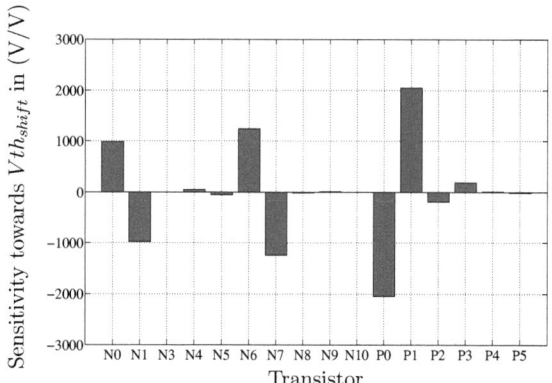

Fig. 4.9: Sensitivity of output referred offset towards aging induced Vth_{shift} of different transistors in the open loop folded cascode OTA circuit

mechanisms to overall folded cascode OTA circuit performance degradation are analyzed and evaluated.

The insight gained from the analytical evaluation of aging degradation mechanisms involved in aging of a folded cascode OTA circuit under asymmetrical stress is discussed next. It can be observed from figure 4.9 that aging induced output referred offset is most sensitive to mismatch in V_{th} shift of the pMOSFET pair P_0 and P_1 in the OTA circuit. The good news is that NBTI degradation of these pMOSFET devices is equal since they see the same gate to source voltage from common mode (cm) feedback circuit and hence compensate each other due to differential signaling. Analysis of data from aging simulation in figure 4.11 and figure 4.12 illustrates that V_{th} shift is a combined effect of

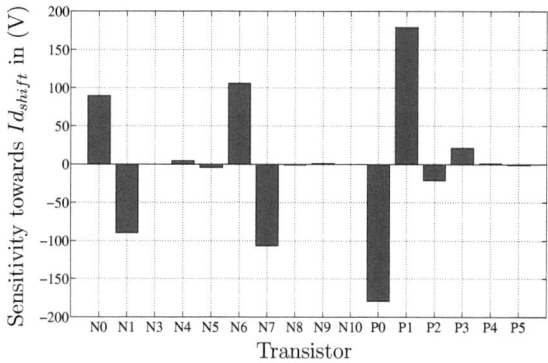

Fig. 4.10: Sensitivity of output referred offset towards aging induced Id_{shift} of different transistors in the open loop folded cascode OTA circuit

Comparison Between Aging of Different OTA Topologies

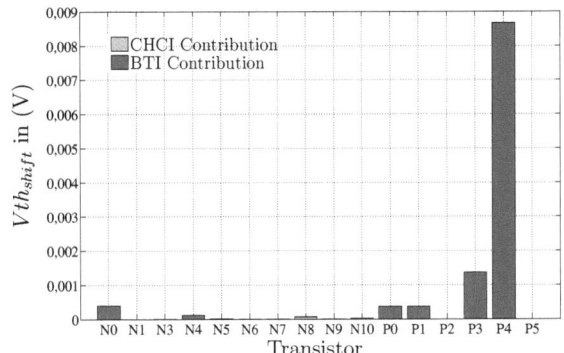

Fig. 4.11: Transistor level Vth_{shift} resulting in the open loop folded cascode OTA circuit after aging simulation with stress time of 4 years at $V_{DD} = 1.155V$ and $T = 85°C$

BTI and CHCI degradation which is evaluated using (2.3), and the I_d shift is a combined effect of CHCI and NCHCI which is evaluated using (2.4). Further output transistor P_4 experiences most V_{th} shift due to high NBTI stress and transistor P_5 experiences most I_d shift due to high NCHCI stress. However, figure 4.13 shows that the dominant contribution to V_{os} due to V_{th} shift comes from input transistor N_0 due to high sensitivity of V_{os} to mismatch in input transistor pair. Similarly, figure 4.14 shows that the dominant contribution to V_{os} due to I_d shift comes from NCHCI degradation of pMOSFET device P_5. Further, it is observed that V_{th} shift makes the dominant contribution (99%) to V_{os}

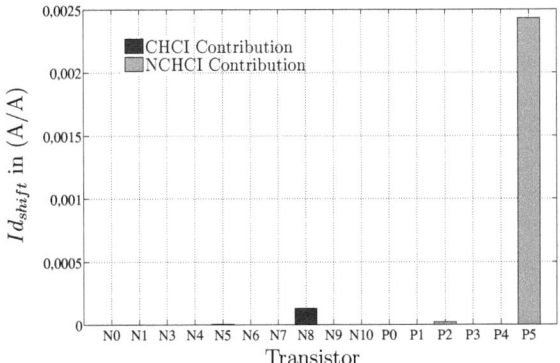

Fig. 4.12: Transistor level Id_{shift} resulting in the open loop folded cascode OTA circuit after aging simulation with stress time of 4 years at $V_{DD} = 1.155V$ and $T = 85°C$

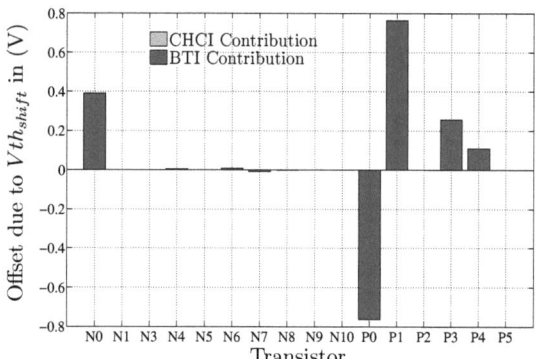

Fig. 4.13: Output referred offset resulting in the open loop folded cascode OTA circuit due to transistor level Vth_{shift} after aging simulation with stress time of 4 years at $V_{DD} = 1.155V$ and $T = 85°C$

compared to I_d shift (1%).

Comparing the simulated results in table 4.2 and table 4.5, it can be noted that under similar asymmetrical stress conditions simple Miller OTA topology is more affected by aging in terms of induced offset compared to folded cascode OTA topology. This results due to a main advantage of folded cascode OTA over simple Miller OTA topology i.e. shielding of the transistors from high bias voltages by the cascode structures. In the cascode structure degradation due to CHCI stress is significantly reduced since the transistors are exposed to reduced drain to source voltages. This finding highlights that

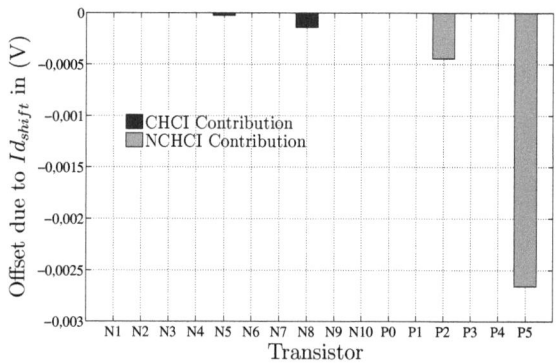

Fig. 4.14: Output referred offset resulting in the open loop folded cascode OTA circuit due to transistor level Id_{shift} after aging simulation with stress time of 4 years at $V_{DD} = 1.155V$ and $T = 85°C$

Stress Condition	Using Simulation	Using Analytical Method
4Yrs, 85°C, $V_{DD} = 1.155V$	0.775	0.784

Table 4.6: Simulated aging induced input referred offset in (mV) resulting in the open loop folded cascode OTA circuit after a stress time of 4Yrs at 85° evaluated using analytical methodology

aging induced performance degradation in OTA is dependent on the circuit topology and analytical evaluation is necessary to locate the weak spots concerning reliability.

4.3 Summary

In this chapter aging induced performance degradation in fully differential OTA circuits was evaluated and discussed. The OTA circuits were designed in 32nm high-κ, metal gate CMOS technology. The investigations were carried out under asymmetrical input stress condition since it induces maximum degradation in matched differential structures. The most degraded circuit performance of OTA circuit under asymmetrical stress was offset. The other performances of the OTA circuit were not much affected, provided all the transistors remained in saturation region.

Aging of open loop and closed loop OTA configurations under similar stress conditions were compared. In closed loop configuration the output transistors which can see full supply voltage levels, experienced most parameter shifts while in open loop condition the input transistor pair witnessed large mismatch due to aging induced parameter shifts. Since the sensitivity of aging induced offset is high towards mismatch in input differential transistor pair, open loop OTA configurations were significantly affected by aging in terms of induced offset compared to closed loop configuration.

Next, two OTA circuit topologies viz., simple Miller OTA and folded cascode OTA both in open loop configuration, were compared for aging induced performance degradation. Performance of folded cascode OTA topology was found to be more robust to aging degradation over simple Miller OTA topology because of the shielding of its transistors from high bias voltages by the cascode structures. Hence under similar asymmetrical stress conditions simple Miller OTA topology was more affected by aging in terms of induced offset compared to folded cascode OTA topology. The importance of circuit topology selection and analytical evaluation to locate the weak spots concerning reliability was thus highlighted.

Chapter 5

Active Countermeasures against Aging Degradation

In chapter 4 it is discussed how the degradation due to aging mechanisms induces the parameter drifts in transistors and results into performance degradation in the closed and open loop operational transconductance amplifier (OTA) circuits. The most degraded circuit performance of OTA circuit under asymmetrical stress is offset. The other performances are not much affected, provided all the transistors remain in saturation region even after aging induced parameter drifts. Aging under asymmetrical stress conditions induces large offset in open loop configuration and small offset in closed loop configuration. In closed loop configuration the input transistors of the OTA circuit see smaller stress compared to the output stage transistors and these output transistors are the main contributors to offset. On the other hand in the open loop configuration the input transistors see a large stress and due to high sensitivity of offset towards these input transistors pair, they are the main contributors to aging induced offset. High resolution mixed signal circuits require OTA circuits with high precision. For e.g. 12-bit data converter with 1V signal range has a smallest resolution of $244\mu V$ and the aging induced offset discussed in section 4.1.2 can introduce errors in its transfer characteristics.

In [5, 23] passive techniques to treat aging induced offset using burn-in and calibration are explained. This type of calibration is off-line and requires re-calibration at regular intervals. Adaptive body biasing technique to compensate aging induced parameter drift is treated in [64]. However it has a drawback of increase in chip area. This chapter introduces active countermeasures to overcome aging induced offset [65]. The two techniques explained in this chapter are Chopper Stabilization (CHS) and Auto Zeroing (AZ) [66,67]. CHS technique tries to preserve the matching of matched transistor pairs by symmetrical degradation of both transistors in the pair and hence it ensures that no offset is induced due to aging. This technique is suitable for all fully differential configurations. On the other hand AZ technique treats the offset by sampling and cancellation, and ensures that the circuit operation is not affected. In this chapter these well known techniques to eliminate DC offset and low frequency (1/f) noise are demonstrated to significantly reduce the effects of performance degradation induced by aging mechanisms in analog circuits.

5.1 Chopper Stabilization

The Chopper Stabilization (CHS) technique is better described as offset stabilization in OTA by using a chopper circuit. This technique has been used for many years to reduce low frequency (1/f) noise and process variation induced DC offset in amplifier circuits. Typically CHS technique is implemented in linear circuits like closed loop amplifier for high precision applications. The application of CHS technique to counteract aging induced performance degradation in fully differential circuits is demonstrated in this section.

5.1.1 Introduction to CHS Technique

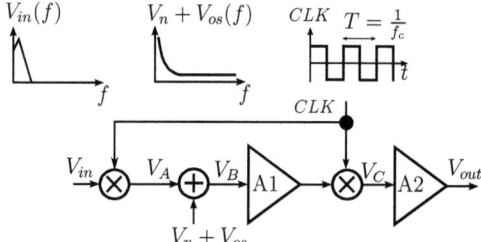

Fig. 5.1: Chopper Stabilization concept

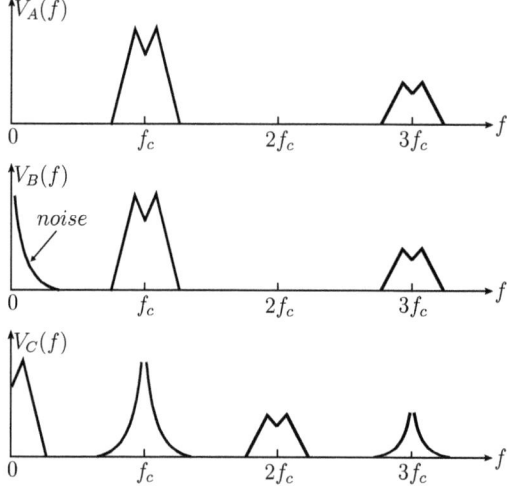

Fig. 5.2: Spectrum at different nodes of the chopper stabilized circuit

Chopper Stabilization

The classical CHS technique is illustrated in figure 5.1 [68]. A CHS circuit consist of an amplifier with two stages where first stage gain (A_1) should be high for effective noise reduction. The two multipliers at the input and output of the first amplifier stage are driven by chopping clock signal (CLK) with frequency (f_c) and amplitude $+1$ and -1. After modulation by the first multiplier, at V_A, the input signal spectrum is shifted to the odd harmonic frequencies of CLK. At the input of the amplifier's first stage, V_B, the undesired noise signals are added to the input signal spectrum. After demodulation by the second multiplier, at V_C, the input signal spectrum is shifted back to its original position and the undesired noise signal spectrum is now shifted to odd harmonic frequencies of CLK. Figure 5.2 shows the spectra at the different nodes to visualize how the DC and low frequency amplifier noise is shifted to higher frequencies outside baseband. Thus, this technique up-converts the input signals to higher frequency using an input multiplier, where the signal is multiplied with a high frequency clock signal. The output signal is again multiplied with the clock signal to convert it back to the baseband. Thus, low frequency noise and offset of the amplifier is eliminated from the output signal. High gain of the amplifier first stage and high chopping frequency enhance the noise suppression.

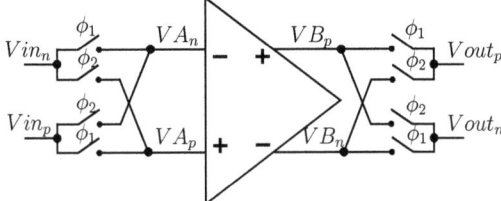

Fig. 5.3: Schematic diagram of chopper stabilized amplifier

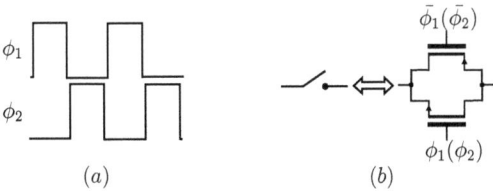

Fig. 5.4: Non-overlapping clock and switch topology for CHS and AZ implementation

Figure 5.3 and 5.4 illustrates the schematic implementation of CHS technique. Here the multipliers are implemented using four pairs of cross coupled switches controlled by non-overlapping clocks. The switches are implemented using CMOS transfer gates, also called transmission gate topology. The second amplifier stage is not shown. During phase when ϕ_1 is "ON" and ϕ_2 is "OFF", Vin_n and Vin_p are connected to VA_n and VA_p respectively and VB_p and VB_n are connected to $Vout_p$ and $Vout_n$ respectively. Whereas during phase when ϕ_1 is OFF and ϕ_2 is ON, Vin_n and Vin_p are connected to VA_p and VA_n respectively and VB_p and VB_n are connected to $Vout_n$ and $Vout_p$

respectively. Due to this modulation at the input and demodulation at the output the average equivalent input noise (Vn_{eq}) of the amplifier first stage is removed, which is given by equation 5.1 [68].

$$Vn_{eq}(\phi_1) = Vn_1 + \frac{Vn_2}{A_1}$$
$$Vn_{eq}(\phi_2) = -Vn_1 + \frac{Vn_2}{A_1}$$
$$Vn_{eq}(average) = \frac{Vn_{eq}(\phi_1) + Vn_{eq}(\phi_2)}{2} = \frac{Vn_2}{A_1} \quad (5.1)$$

where Vn_1 and Vn_2 are the input noise of the amplifier first and second stage respectively and A_1 is the gain of the amplifier first stage.

5.1.2 Reduction in Aging Degradation using CHS Technique

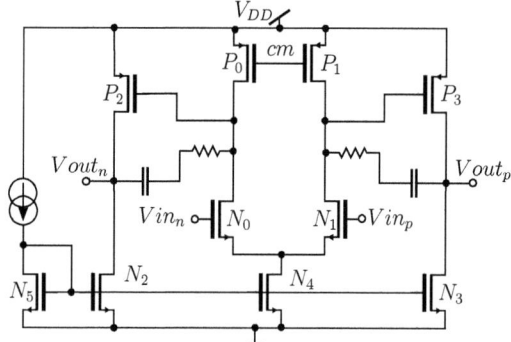

Fig. 5.5: Schematic of simple Miller OTA circuit

The use of CHS technique in fully differential structures causes both transistors of the input differential pair and other matched pairs in the circuit to be stressed equally due to the continuous switching of the input connection via the cross coupled switches which is quite beneficial with respect to aging. This equal distribution of stress in matched pair's results into negligible mismatch and ideally zero offset after aging since the degradation

Stress Condition	Without CHS	With CHS
4Yrs, 85°C, $V_{DD} = 1.155V$	5.29	2.223E-6
10^3s, 125°C, $V_{DD} = 1.394V$	6.099	9.611E-7

Table 5.1: Simulated input referred offset in (mV) resulting in the aged simple Miller OTA circuit with and without CHS technique evaluated at $T = 25°C$ and $V_{DD} = 1V$

Chopper Stabilization

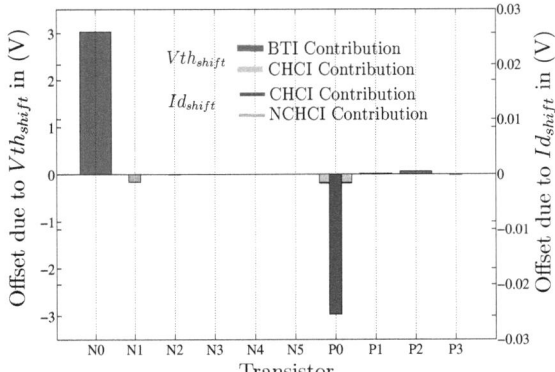

Fig. 5.6: Output referred offset resulting in open loop OTA circuit without CHS due to transistor level Vth_{shift} and Id_{shift} after aging simulation with stress time of 10^3s at $V_{DD} = 1.394V$ and $T = 125°C$

effects cancel one another in the differential signal. The comparison of aging related performance degradation of the simple Miller OTA circuit illustrated in figure 5.5 (from section 4.2.1) with and without CHS technique, for the mobile phone EoL use case of 4 years at 85°C with 105% of worst case V_{DD}, ($V_{DD} = 1.155V$), under asymmetrical DC input stress ($Vin_n = 1.155V$, $Vin_p = 0V$) is presented next using aging simulation and test chip measurements. The equivalent mapped accelerated aging bias condition for stress time of 10^3s at 125°C was derived to be $V_{DD} = 1.394V$ as discussed in section 3.2. Therefore for measurement purposes the OTA was stressed with static (DC) asymmetrical

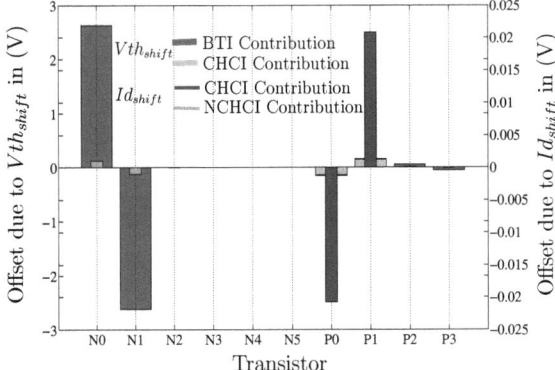

Fig. 5.7: Output referred offset resulting in open loop OTA circuit with CHS due to transistor level Vth_{shift} and Id_{shift} after aging simulation with stress time of 10^3s at $V_{DD} = 1.394V$ and $T = 125°C$

48   Chapter 5. Active Countermeasures against Aging Degradation

stress of $V_{DD} = 1.394V$, $Vin_p = 0V$ and $Vin_n = 1.394V$ at 125°C for 10^3s.

CHS technique results in symmetrical degradation of the transistor pairs in the OTA circuit as is confirmed by the aging simulation results. From figure 5.6 it can be observed that in the OTA circuit without CHS, application of asymmetric stress induces mismatch in matched pairs due to degradation of differential transistor pairs which leads to generation of offset. Whereas, figure 5.7 clearly shows that in the OTA circuit with CHS, continuous switching of input stress causes symmetrical degradation of all transistor pairs and hence offset cancellation. With CHS technique the stress time for each transistor in the circuit is reduced to half, however only a small reduction in BTI induced Vth_{shift} ($\approx 12\%$) is observed in figure 5.7 compared to that in figure 5.6. This results due to the BTI degradation's power law behavior in time ($\Delta V_{th} \propto t^n$), where n is around 0.19 to 0.2. The simulation results in figure 5.7 do not account for the AC factor due to the recovery effect which can reduce the BTI degradation by a factor of two [49]. Degradation of the switches results into their V_{th} shift over lifetime as they are operated with full swing clock signals, however this does not affect the circuit performance at moderate operating frequencies. Table 5.1 compares the simulated degradation induced input referred offset in simple Miller OTA circuit without and with CHS technique evaluated at $T = 25°C$ and $V_{DD} = 1V$.

Simulation results show that the induced input referred offset is reduced by more than 99% due to the inherently equal distribution of stress on the input transistors and other matched transistor pairs while using the chopper stabilization circuitry.

5.1.3   Measurements

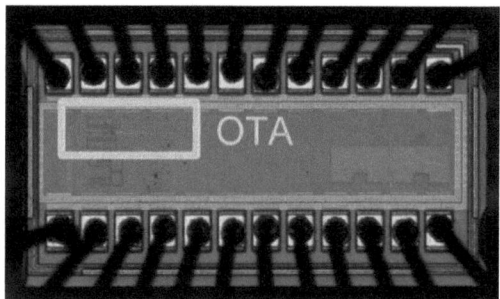

Fig. 5.8: Die photograph of the measured 32nm HK/MG simple Miller OTA test chip

To prove the concept of mitigation of aging mechanisms induced mismatch and offset using CHS technique, measurements are performed on simple Miller OTA test chips fabricated using state-of-the-art 32nm high-κ (HK), metal gate(MG) CMOS technology. The schematic of this implemented OTA is illustrated in figure 5.5. The die photograph of the measured test chip is illustrated in figure 5.8. The measurements are performed

Chopper Stabilization

Fig. 5.9: Measurement test setup for accelerated aging of simple Miller OTA test chip

on four test chip samples (S1-S4). Samples S1-S2 are used to measure the aging induced performance degradation (offset) without, while S3-S4 are used to measure the offset with CHS technique. The virgin (before stress) performances of S1-S4 are presented in table 5.2. Accelerated aging conditions similar to that used during simulations i.e. static (DC) asymmetrical stress of $V_{DD} = 1.394V$, $Vin_p = 0V$ and $Vin_n = 1.394V$ at 125°C for 10^3s, are applied to the samples. For the CHS technique, chopping clock frequency (f_c) of 5Hz is used for S3 and 500Hz is used for S4. In this section the aging measurement setup, its limitations and results with and without CHS technique are presented.

The CHS technique is implemented using discrete components around the test chip. Since the switches are selected to withstand high voltages and were implemented externally, they are expected to experience negligible aging degradation. The entire measurement activity is divided into following five steps and is repeated for all samples:

$$M1 \Rightarrow M2 \Rightarrow S \Rightarrow M3 \Rightarrow A \Rightarrow M4 \Rightarrow M5$$

where,

"S" → Stress phase with stress time of 10^3s at 125°C and $V_{DD} = 1.394V$
"A" → Annealing phase at high temperature, 125°C and $V_{DD} = 0V$ for 10^4s to study

Performance	Simulated	Measured			
		S1	S2	S3	S4
DC Gain (dB)	53.43	59.20	60.51	57.84	60.73
V_{os} (mV)	0	210	420	691	230

Table 5.2: Simulated and measured simple Miller OTA DC gain and output referred offset before stress evaluated at $T = 25°C$ and $V_{DD} = 1V$

Chapter 5. Active Countermeasures against Aging Degradation

Measurement Step	Without CHS		With CHS	
	S1	S2	S3	S4
M2, before aging	0.23	0.34	0.769	0.147
M4, after annealing	−2.71	−2.47	0.758	0.248
$\Delta offset$	−2.94	−2.81	−0.011	0.101

Table 5.3: Measured input referred offset (in mV) resulting in the aged simple Miller OTA circuit with and without CHS technique evaluated at 125°C

long term permanent offset behavior

And the offset measurements were performed,
"M1" → before stress at 25°C and $V_{DD} = 1V$
"M2" → before stress at 125°C and $V_{DD} = 1V$
"M3" → post stress relaxation at 125°C and $V_{DD} = 1V$
"M4" → post annealing at 125°C and $V_{DD} = 1V$
"M5" → post annealing at 25°C and $V_{DD} = 1V$.

In the accelerated measurement test setup, the temperatures during measurement, stress and annealing phases are precisely controlled by monitoring temperature of the Peltier element which is attached to the top casing of the test chip as illustrated in figure 5.9. The measurement setup is completely automated using LabVIEW software. Table 5.3 compares the aging degradation induced input referred offset in simple Miller OTA test chips without and with CHS technique measured at $T = 125°C$ and $V_{DD} = 1V$ before aging and after annealing.

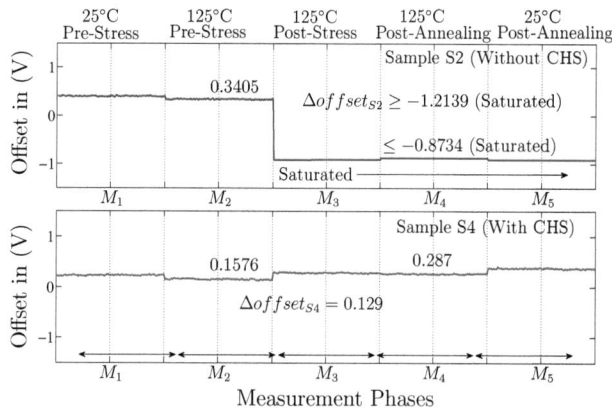

Fig. 5.10: Measured output referred offset transition plots in simple Miller OTA during different measurement phases for samples S2 and S4, without and with CHS respectively

Chopper Stabilization 51

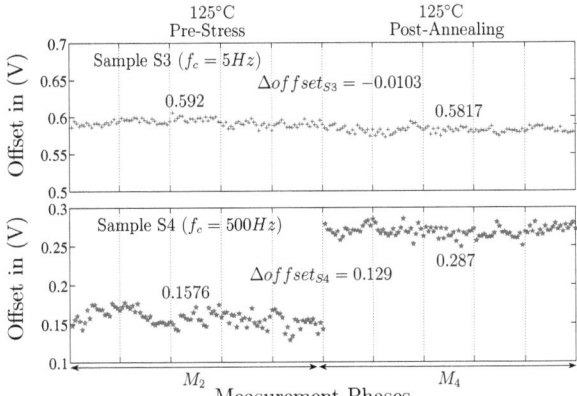

Fig. 5.11: Measured output referred offset transition plots in simple Miller OTA during pre stress and post annealing phases for samples S3 (f_c = 5Hz) and S4 (f_c = 500Hz) with CHS technique

$$\Delta offset = offset_{post\ annealing} - offset_{pre\ stress} \quad (5.2)$$

Figure 5.10 illustrates the transition plots of measured output referred offset in simple Miller OTA test chips during different phases of measurement for samples S2 and S4, without and with CHS technique respectively. Here $\Delta offset$ is evaluated at 125°C using (5.2). Comparing the aging induced offset ($\Delta offset$) values in the OTA circuit without and with CHS technique it is proved that use of CHS techniques results into mitigation of the permanent component of aging degradation. The $\Delta offset$ is reduced by around 96% for sample S4, when compared to S2. It is also shown that with the use of CHS technique the relaxation of offset post stress is not visible. Negligible relaxation of offset occurs after stress phase and post annealing, as observed from the offset transition graph for S4. Aging of S2 generated large offset value. It is known from [3] that accelerated aging without CHS leads to large relaxation of this aging induced offset. But this could not be observed here due to saturation effects. Due to open loop configuration and high gain of the OTA circuit, the output voltages are saturated for large values of aging induced offset during both post stress and annealing phases. So for these saturated output voltages, the relaxation behavior and the exact values of the generated output referred offset could not be measured. With CHS technique the aging induced offset is very small hence no saturation is observed and also the relaxation parts are compensated. The input referred offset in table 5.3 is evaluated before and after stress by connecting one of the OTA circuit inputs to a bias level of V_{cm} and sweeping the voltage at the other input till the output referred offset is equal to 0.

Figure 5.11 illustrates the plot of measured transition of output referred offset in simple Miller OTA test chips during two phases of measurement (before stress and after

52 Chapter 5. Active Countermeasures against Aging Degradation

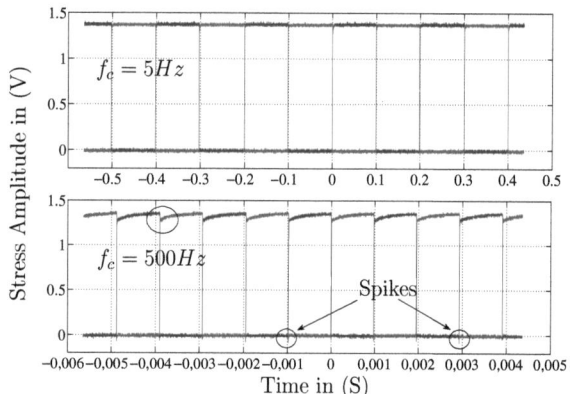

Fig. 5.12: Measured stress voltages generated at the gate of input differential pair transistors of the simple Miller OTA circuit while using CHS technique

annealing) at 125°C for samples S3 and S4, both with CHS technique but operated at different chopping frequencies. Sample S3 is modulated and demodulated at 5Hz and S4 at 500Hz. The $\Delta offset$ is reduced by around 99% for sample S3 with low chopping frequency and around 96% for S4 with high chopping frequency, when compared to S1 in table 5.3. The somewhat higher value of $\Delta offset$ for S4 results due to the limitation of the existing measurement setup which is built for DC and low frequency measurements. As illustrated in figure 5.12, the static (DC) asymmetrical stress applied at the input of the CHS circuit turns into a toggling waveform before reaching the input transistor pair of the OTA circuit, due to cross coupled input switches of the CHS. It can be observed that waveform at the frequency of 5Hz was much cleaner and symmetric compared to that at 500Hz where spikes and loading occurs. The aging induced degradation is highly sensitive to the exact value of the stress voltages applied at the input transistors due to the high gain of the circuit and exponential dependence seen in (2.1). Therefore, the measurement could be performed more precisely with integrated switches. Nevertheless, significant reduction in aging degradation using the chopping technique concept could be proved experimentally in both cases.

In this investigation mismatch induced due to process variation and variation in aging degradation is not considered. Device variations due to technology fluctuations are known to cause parameter variations, e.g. threshold voltage mismatch due to statistical variation of the number of doping atoms in a transistor. In the same way, also degradation effects show mismatch, e.g. due to varying number of traps in the oxide and the interface to the channel. Such statistical effects are not included in our simulation models, and they could counteract the offset compensation obtained in the CHS technique for circuits implemented using minimum size transistors. However the input transistors of an amplifier have much larger area, so there is much less mismatch of the degradation behavior due to averaging effects as reported in [5]. This can also be seen from the measurement

Auto Zeroing

results in table 5.3, by the very small aging induced $\Delta offset$ in S3.

CHS techniques when implemented for a closed loop OTA configuration, can additionally cancel offset due to process variation and low frequency flicker (1/f) noise as explained in section 5.1.1. This makes the circuit robust over entire operation lifetime.

5.2 Auto Zeroing

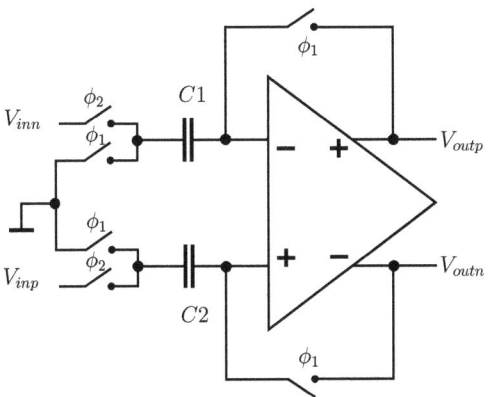

Fig. 5.13: Schematic diagram of comparator with auto zeroing

The classical Auto Zeroing (AZ) technique which is sometimes also referred to as correlated double sampling is illustrated in figure 5.13 [69]. This technique has been used for many years to reduce low frequency flicker (1/f) noise and DC offset in amplifier circuits. This technique is applicable for sampled data systems since the input signals are disconnected from the circuit during the sampling phase.

5.2.1 Introduction to AZ Technique

The basic principle of Auto Zeroing (AZ) technique is to sample the DC offset and low frequency noise using switched capacitor circuits and then subtract it from the input signal of the OTA. This technique is suitable for sampled data systems and requires minimum two non-overlapping clock phases for offset cancellation. During the sampling phase (ϕ_1 is ON and ϕ_2 is OFF) the DC offset and the low frequency noise are sampled and stored on the capacitors. Next, during the cancellation or zeroing phase (ϕ_1 is OFF and ϕ_2 is ON) these stored values are subtracted from the input signal. Thus the sampled unwanted offset and noise on the capacitors lead to cancellation. The two phases of non-overlapping clocks and the switches implemented using transmission gate topology are illustrated figure 5.4.

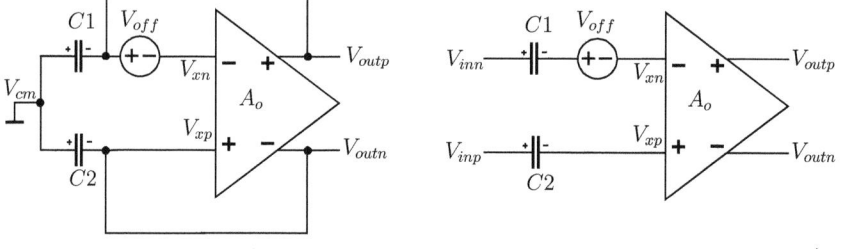

Sampling phase (ϕ_1 ON) ($t = nT$) Compensation phase (ϕ_2 ON)($t = (n + \frac{1}{2})T$)

Fig. 5.14: Schematic diagram of comparator with AZ

The circuit configuration during two phases of AZ technique implemented for a fully differential simple Miller OTA topology is illustrated in figure 5.14. The circuit operation is explained as follows: during sampling phase when ϕ_1 is ON and ϕ_2 is OFF, ideally the differential input voltage of the OTA is V_{off} and this is sampled on the capacitor. Next during the compensation phase when ϕ_2 is ON and ϕ_1 is OFF, the differential input voltage of the OTA is given by $V_{inp} - V_{inn} + V_{off} - V_{off} = V_{inp} - V_{inn}$. The accuracy of the offset cancellation using AZ technique depends on the open loop DC gain of the OTA (A_o).

For ideal OTA (where A_o is the open loop DC gain of the OTA)

$$(V_{outp} - V_{outn}) = A_o \cdot (V_{xp} - V_{xn}) \tag{5.3}$$

And for the OTA in a closed loop configuration (where $A_o \gg 1$):

$$V_{xp} \approx V_{xn}$$

The schematic on the left in figure 5.14 shows the circuit configuration during sampling phase (ϕ_1 ON). The OTA is operated in a unity gain feedback loop, thus the voltages at the output terminals at time ($t = nT$) are given by:

$$V_{outp}[nT] = V_{xn}[nT] + V_{off}$$
$$V_{outn}[nT] = V_{xp}[nT]$$
$$\Rightarrow (V_{outp} - V_{outn})[nT] = V_{xn}[nT] + V_{off} - V_{xp}[nT] \tag{5.4}$$

Substituting $(V_{outp} - V_{outn})$ in equation (5.4) with result in equation (5.3) we have,

$$A_o \cdot (V_{xp} - V_{xn})[nT] = V_{xn}[nT] + V_{off} - V_{xp}[nT]$$
$$\Rightarrow V_{off} = (A_o + 1) \cdot (V_{xp} - V_{xn})[nT]$$
$$\Rightarrow (V_{outp} - V_{outn})[nT] = \frac{A_o}{A_o + 1} \cdot V_{off} \tag{5.5}$$

Auto Zeroing

$$V_{C1}[nT] = V_{cm} - V_{outp}[nT] \quad (5.6)$$
$$V_{C2}[nT] = V_{cm} - V_{outn}[nT] \quad (5.7)$$

Next as shown by the schematic on the right in figure 5.14, during the compensation phase (ϕ_2 ON) the OTA is operated in open loop configuration and now the voltages at the input terminals at time ($t = (n + \frac{1}{2})T$) are given by:

$$V_{xn}[(n + \frac{1}{2})T] = V_{inn}[(n + \frac{1}{2})T] - V_{C1}[nT] - V_{off}$$
$$V_{xp}[(n + \frac{1}{2})T] = V_{inp}[(n + \frac{1}{2})T] - V_{C2}[nT]$$

Substituting $V_{C1}[nT]$ and $V_{C2}[nT]$ with result in equation (5.6) and (5.7) we have,

$$(V_{xp} - V_{xn})[(n + \frac{1}{2})T] = (V_{inp} - V_{inn})[(n + \frac{1}{2})T] - (V_{outp} - V_{outn})[nT] + V_{off} \quad (5.8)$$

Substituting $(V_{outp} - V_{outn})[nT]$ in equation (5.8) with result in equation (5.5) we have,

$$(V_{outp} - V_{outn})[(n + \frac{1}{2})T] = A_o \cdot ((V_{inp} - V_{inn})[(n + \frac{1}{2})T] + \underbrace{\frac{1}{A_o + 1} \cdot V_{off}}_{\text{Residual Offset Error}}) \quad (5.9)$$

The expression under curly bracket is termed as residual offset error which results from finite OTA gain. The error induced by charge injection from switches is assumed to be zero due to full differential configuration. Thus the above analysis shows that the accuracy of offset compensation using the AZ technique is dependent on the open loop gain of the amplifier circuit.

5.2.2 Reduction in Aging Degradation using AZ Technique

The same OTA circuit as illustrated in figure 5.5 is simulated with AZ technique for the mobile phone EoL use case of 4 years at 85°C with 105% of worst case V_{DD} ($V_{DD} = 1.155V$), under asymmetrical DC input stress ($Vin_n = 1.155V$, $Vin_p = 0V$) and for the equivalent mapped accelerated aging bias condition for stress time of 10^3s at 125°C with $V_{DD} = 1.394V$, $Vin_n = 1.394V$ and $Vin_p = 0V$ using a clock frequency of 12.5MHz. As illustrated in figure 5.15, the degradation effects in the simple Miller OTA circuit using AZ do not compensate each other as compared to the one using CHS. The order of relevance of the degradation mechanisms and the induced mismatch in matched transistor pairs remains the same as that of figure 5.6 without AZ. However, since the OTA is alternately operated in closed and open loop configurations, which results from the two operation phases, the total time when stress is applied to the input differential pair is reduced by one half. Therefore, the absolute values of the corresponding offset components are reduced, and the comparison in table 5.4 shows that still significant offset remains due to aging. This is the offset of the amplifier by itself induced due to aging. When the offset

56　　　　　　　　　Chapter 5. Active Countermeasures against Aging Degradation

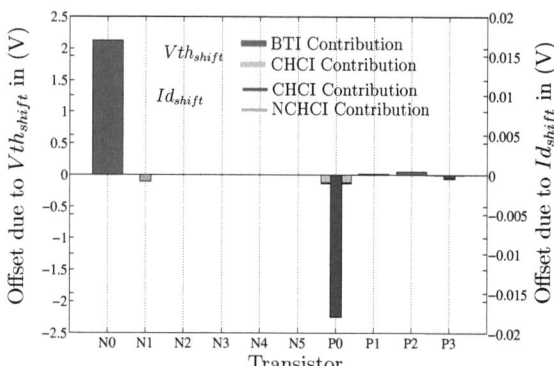

Fig. 5.15: Output referred offset resulting in open loop simple Miller OTA circuit with AZ due to transistor level Vth_{shift} and Id_{shift} after aging simulation with stress time of 10^3s at $V_{DD} = 1.394V$ and $T = 125°C$

of the comparator is regarded after applying the autozeroing for offset cancellation, it looks much better (99% reduction), see last column of table 5.4. The storage of the offset on the capacitor and cancellation during the next phase is able to remove almost all the degradation induced offset. The remaining offset is the residue offset error resulting from finite OTA gain as discussed in section 5.2.1. Similar to the circuit with CHS technique, the degradation of the switches results in increase of their ON resistance but does not affect the circuit performance at moderate operating frequencies.

In this section, only the permanent part of degradation is simulated and analyzed, mainly due to the lack of a model for the recovery parts of degradation. But the discussed technique can also be beneficial for these effects. AZ technique is known to reduce effects of low frequency 1/f noise. The low frequency part of the recovery component of the degradation is similar to 1/f noise. Therefore, recovery components with time constants larger than the clock period can be compensated in the same way as the slowly varying permanent stress effects. Thus this technique also has a potential to reduce performance deterioration due to the recovery component of the stress degradation.

Stress Condition	Without AZ	With AZ (due to aging)	With AZ (after offset cancellation)
4Yrs, 85°C, $V_{DD} = 1.155V$	5.29	3.08	1.014E-3
10^3s, 125°C, $V_{DD} = 1.394V$	6.099	4.31	9.161E-3

Table 5.4: Simulated input referred offset in (mV) resulting in the aged OTA circuit with and without AZ technique evaluated at $T = 25°C$ and $V_{DD} = 1V$

5.3 Summary

In this chapter two solutions, chopper stabilization and auto zeroing to reduce the effects of aging degradation in analog and mixed signal circuits were proposed and evaluated.

Chopper stabilization is usually used in analog circuits to eliminate offset and low frequency noise. In this chapter its application to mitigate aging related performance degradation in fully differential circuits was proposed. This mitigation of aging degradation results due to symmetrical degradation of the matched transistor pairs. No mismatch in matched transistor pairs results into zero offset and cancellation of aging induced parameter drifts due to differential signaling. The concept was proved with measurement results on the test chips fabricated using 32nm high-κ, metal gate CMOS technology. Using CHS aging induced performance degradation in the implemented fully differential OTA was reduced by more than 96% (measured). It was also shown that using CHS, the relaxation of the offset due to the BTI mechanism was not visible. Hence dynamic errors resulting from this relaxation behavior could be avoided. In the examined samples, there was no mismatch in degradation induced parameter drifts which was to be expected in general due to large analog size input transistors. Thus the concept of CHS could be effectively used to eliminate mismatch induced due to aging in all differential circuits.

Auto zeroing is another technique used in sampled data circuits to eliminate offset and low frequency noise. With AZ, the aging degradation induced offset is stored on the input capacitors and canceled during the next clock phase, resulting in around 99% reduced offset (simulated) during circuit operation. AZ is suitable for sampled data systems, since the input signals are disconnected from the circuit during the offset sampling phase. However the accuracy of offset compensation using the AZ is dependent on the open loop gain of the amplifier circuit. And a low open loop gain results into residue offset errors. On the other hand, mitigation of the aging degradation using CHS is independent of the amplifier open loop gain. And it can be used in both continuous time and sampled data systems. AZ also has the potential of reducing the low frequency components of relaxing degradation in amplifiers circuits.

Chapter 6
Aging in Ring Oscillator Circuits

This chapter provides evaluation of aging degradation in ring oscillator circuit typically used as voltage controlled oscillator (VCO) for clock generation, which is one of the important building blocks in mixed signal applications. Ring oscillator is also often used to demonstrate and evaluate new technology nodes. Investigations related to the reliability of the ring oscillator circuits implemented in CMOS technology are presented in [49,51,70,71]. A model to predict time dependence of ring oscillator aging degradation is presented in [72]. It is reported that aging of the ring oscillator leads to reduction in the supply current (I_{DD}) and degradation in the oscillator switching frequency (f_{osc}). Further increase in lifetime was observed under dynamic (AC) stress compared to static (DC) stress considering bias temperature instability as the main contributor to ring oscillator performance degradation. However, due to aggressive non-constant field scaling, hot carrier injection is again a prime concern for device and circuit reliability [57].

Hence, in this chapter the contributions of different aging mechanisms viz. NBTI, PBTI, CHCI and NCHCI towards parameter drifts in transistors of the ring oscillator circuit implemented in 32nm high-κ metal gate CMOS technology, stressed under both AC and DC stress are analyzed and discussed. Further a new circuit technique to efficiently monitor and compensate aging induced performance degradation in the ring oscillator circuit is demonstrated. This simple closed loop feedback technique monitors the performance degradation due to aging and compensates it by reducing the resistance of the CMOS switch added at the input terminal of each inverter stage. The switch control voltage tracks the ring oscillator degradation and provides useful information about its aging behavior.

6.1 Ring Oscillator

A typical ring oscillator comprises of an odd number of CMOS inverter stages. The output of each inverter is used as input for the next one. The last output is fed back to the first inverter stage. The frequency (f_{osc}) depends on the number of stages (N) and the

Chapter 6. Aging in Ring Oscillator Circuits

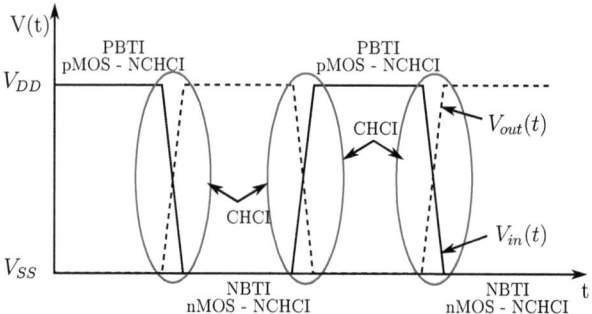

Fig. 6.1: Different aging mechanisms causing degradation in CMOS inverter stage of a ring oscillator circuit under AC stress condition

delay time of each inverter stage (τ_d). The f_{osc} can be approximated using equation (6.1)

$$f_{osc} = \frac{1}{2 \cdot N \cdot \tau_d}$$

$$\tau_d = \frac{\tau_n + \tau_p}{2} \quad (6.1)$$

$$\tau_n = \frac{V_{DD} \cdot C_L}{Id_n}, \tau_p = \frac{V_{DD} \cdot C_L}{Id_p}$$

where τ_n, τ_p are the switching time constant of nMOSFET and pMOSFET transistors. Id_n, Id_p are the drain current of nMOSFET and pMOSFET transistors at $|V_{gs}| = V_{DD}$. And C_L is the load capacitance which is a combination of input oxide and parasitic output capacitance of the inverter stage.

During normal operation of the ring oscillator circuit both the nMOSFET and pMOSFET transistors in the inverter stages are degraded by aging under AC stress condition. Figure 6.1 illustrates input ($V_{in}(t)$) and output ($V_{out}(t)$) waveforms of an inverter stage. As depicted here the stress conditions can be divided into two main regions. When the input is in the steady-state i.e. equal to supply voltage (V_{DD}) or ground (V_{SS}), the "ON" transistors are exposed to BTI and the "OFF" transistors to NCHCI stress. In the region when the input makes a transition, both transistors are exposed to CHCI. In case when the ring oscillator circuit is exposed to only DC stress condition, e.g. during disable or power down mode, all the ON transistors are exposed to BTI and the OFF transistors to NCHCI stress. In the absence of any transition degradation resulting from CHCI stress is missing.

Transistor	W/L [nm/nm]
pMOSFET ($\frac{W}{L}_p$)	1020/30
nMOSFET ($\frac{W}{L}_n$)	690/30

Table 6.1: MOSFET devices W/L ratios

Ring Oscillator

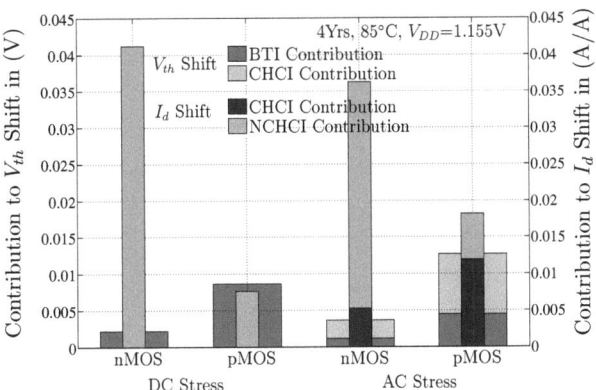

Fig. 6.2: Comparison between contribution of different aging mechanisms causing degradation in nMOSFET and pMOSFET transistors of the ring oscillator circuit both under AC and DC stress for mobile phone EoL use case conditions

The aging related performance degradation of the ring oscillator circuit is evaluated for a mobile phone EoL use case of 4 years at 85°C with 105% of maximum specified V_{DD} ($V_{DD} = 1.155V$), under both AC and DC stress conditions. The equivalent accelerated aging bias condition for stress time of 10^4s at 125°C is derived to be $V_{DD} = 1.481V$. The aging simulation is performed on the ring oscillator circuit consisting 23 inverter stages with (W/L) ratios of the MOSFET devices presented in table 6.1 and imple-

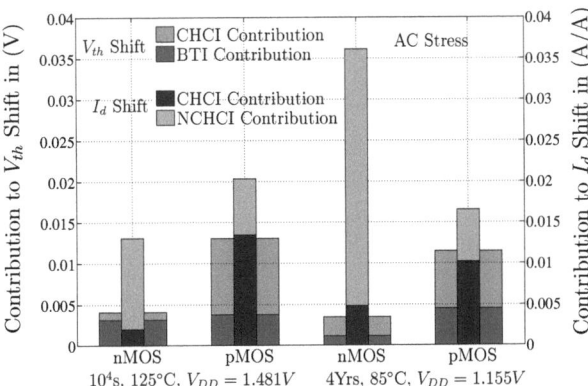

Fig. 6.3: Comparison between contribution of different aging mechanisms causing degradation in nMOSFET and pMOSFET transistors of the ring oscillator circuit under AC stress for mobile phone EoL use case and accelerated conditions

Characterized at → Stress Condition ↓	$V_{DD} = 0.9V$ Simulated	$V_{DD} = 0.9V$ Measured	$V_{DD} = 1.481V$ Simulated
Use case (DC)	1.33	–	0.707
Accelerated (DC)	1.104	1.57	0.394
Use case (AC)	4.06	–	1.75
Accelerated (AC)	3.6	2.3	1.21

Table 6.2: Measured and simulated relative degradation of the ring oscillator frequency (Δf_{osc}) in percentage under DC and AC use case and accelerated stress conditions evaluated at different V_{DD} values for $T = 125°C$

mented using 32nm high-κ metal gate CMOS technology. The simulated f_{osc} before stress is $1.38GHz$, evaluated at $V_{DD} = 1V$ and $T = 125°C$.

The comparison between aging induced parameter drifts in nMOSFET and pMOSFET transistors of the ring oscillator circuit under AC and DC stress conditions is depicted in figure 6.2. It is observed that the degradation due to BTI stress is around factor of two more under DC stress compared to that under AC stress considering AC factor due to the recovery effect. Moreover, parameter shifts due to CHCI and NCHCI degradation under AC stress lead to overall higher performance degradation in the ring oscillator circuit. These results highlight enhanced HCI degradation at 32nm technology node compared to 130nm node discussed in [51] where BTI degradation was the main contributor to ring oscillator performance degradation and lifetime enhancement was found under AC stress compared to that under DC stress.

Figure 6.3 shows the comparison between contribution of different aging mechanisms in nMOSFET and pMOSFET transistors of an inverter stage in the aged ring oscillator under use case and accelerated AC stress conditions. It is observed that the composition of aging mechanisms remains nearly same for the pMOSFET device and in case of nMOSFET device the NCHCI contribution is particularly reduced for lower stress time in accelerated stress condition. However, the frequency degradation in both the conditions is similar. The difficulty in exact mapping of accelerated stress setup to the use case condition arises since more than one dominant aging effect prevails in the ring oscillator circuit.

Table 6.2 presents the simulated and measured degradation of f_{osc} in percentage under DC and AC use case and accelerated stress conditions characterized at different V_{DD} values for $T = 125°C$. In the case of AC stress condition, the results do not account for the AC factor due to the recovery effect. This could be the reason for lower measured value of degradation under accelerated AC stress compared to the corresponding simulation result. It should be noted that under AC stress both transistors of each inverter stage witnesses BTI degradation whereas under DC stress only one of the two transistors in each inverter is degraded by BTI stress. Nevertheless, the analysis presented in figure 6.2 and the measurement results confirm an important result that unlike in previous technology nodes, degradation due to HCI mechanism under AC stress leads to enhanced performance degradation in the ring oscillator circuit compared to BTI degradation and

Aging Monitor and Compensation Circuit

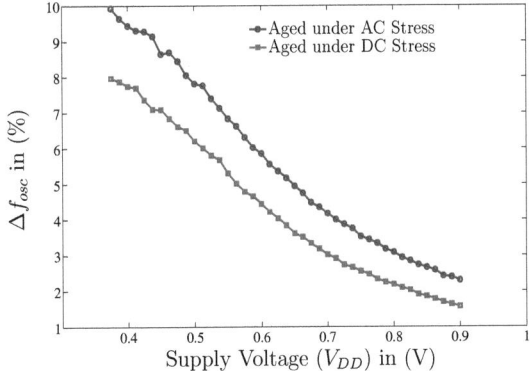

Fig. 6.4: Measured relative degradation of the ring oscillator frequency (Δf_{osc}) in percentage vs. supply voltage (V_{DD}) characterized at $T = 125°C$

hence lifetime enhancement under AC stress is no more guaranteed.

$$\Delta f_{osc}\% = \frac{f_{osc(pre-stress)} - f_{osc(post-stress)}}{f_{osc(pre-stress)}} \cdot 100 \qquad (6.2)$$

Further, the relative performance degradation of the aged ring oscillator circuit increases as the operating voltage (V_{DD}) during characterization decreases due to reduced transistor voltage headroom and overdrive [71]. This is depicted in figure 6.4, where relative degradation of f_{osc} is evaluated using equation (6.2), measured at different V_{DD} values for two test chips stressed under DC and AC accelerated stress conditions respectively. It can be observed that the relative degradation increases as V_{DD} reduces. And the degradation under AC stress is more compared to that under DC stress due to high HCI contribution even with reduction of BTI contribution resulting from the recovery effect. The contribution of CHCI degradation reduces as f_{osc} decreases due to less number of signal transitions.

6.2 Aging Monitor and Compensation Circuit

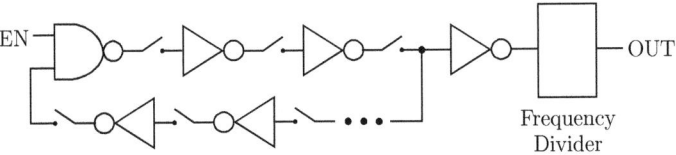

Fig. 6.5: Ring oscillator circuit schematic with aging monitor and compensation

Chapter 6. Aging in Ring Oscillator Circuits

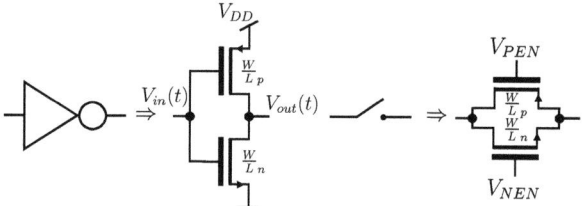

Fig. 6.6: Schematic of CMOS inverter and switch used in the ring oscillator circuit

To monitor the performance degradation in the ring oscillator circuit under various aging mechanisms and to compensate the degradation effects, a special ring oscillator structure is implemented on the test chip in 32nm high-κ metal gate CMOS technology. This modified ring oscillator with aging monitor and compensation circuit technique is illustrated in figure 6.5. Here a CMOS switch stage is added at the input of each inverter stage. The detailed schematics of an inverter and switch stage with the (W/L) ratios of the MOSFET devices are shown in figure 6.6 and table 6.1 respectively. The enable signal (EN) allows the oscillator to be stressed under static DC or dynamic AC state. There are total 23 inverter and switch stages in the circuit. The frequency divider circuit is added to divide the f_{osc} by a frequency divider ratio $(Divider)$ of 2^{10} which makes the measurement setup simple. This ring oscillator circuit modification changes the equation for f_{osc} from equation (6.1) to (6.3).

$$f_{osc} = \frac{1}{2 \cdot N \cdot \tau_d \cdot Divider}$$

$$\tau_d = \frac{\tau_n + \tau_p}{2} + \tau_s \qquad (6.3)$$

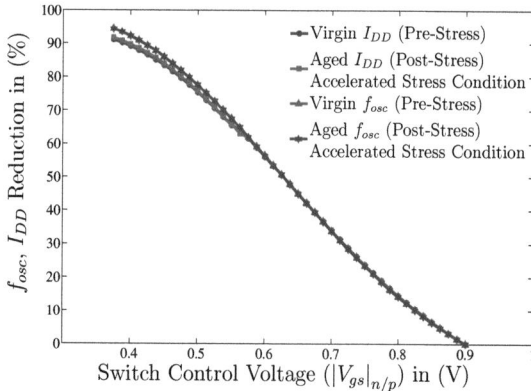

Fig. 6.7: Measured sensitivity of the ring oscillator frequency (f_{osc}) and supply current (I_{DD}) towards the switch control voltage $(|V_{gs}|_{n/p})$ in percentage characterized at $V_{DD} = 0.9V$ and $T = 125°C$

Aging Monitor and Compensation Circuit 65

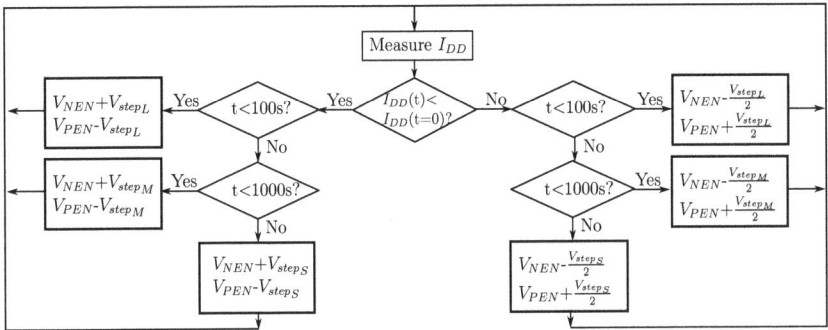

Fig. 6.8: Aging degradation monitor and adaptive bipolar compensation algorithm

$$\tau_s = R_s \cdot C_L$$

where, R_s is the equivalent resistance of the switch stage.

To monitor and compensate the aging degradation, virgin (pre-stress, $t = 0$) performance (I_{DD} or f_{osc}) of the ring oscillator circuit is measured and stored as reference. During the operating life time of the circuit (in this case, mapped accelerated stress time ($= 10^4 s$) equivalent to EoL mobile phone use case condition), aging induced performance degradation is monitored which can be achieved using a sensing hardware. After each monitoring step ($t_{step} = 5s$) the degradation is compensated by controlling R_s, which in turn is varied by nMOSFET switch control voltage (V_{NEN}) and pMOSFET switch control voltage (V_{PEN}), for example using a micro-controller. Therefore this compensation technique works in the background without affecting the normal circuit operation. The measured sensitivity of I_{DD} to change in the switch control voltages ($|V_{gs}|_{n/p}$) is illustrated in figure 6.7. It can be observed that the ring oscillator has a linear and wide tuning range over the switch control voltages. Also the switch control voltages are not elevated during stress so there is no degradation in switch performance during and after stress.

Aging results into weakening of nMOSFET and pMOSFET transistors in the ring oscillator circuit over stress time, hence it causes reduction in I_{DD}. In order to compensate this degradation, based on the sensitivity of I_{DD} towards switch control voltage, we defined a small voltage step ($V_{step} = 50\mu V$) to be added to V_{NEN} and subtracted from V_{PEN} in case the monitored I_{DD} is smaller than the reference I_{DD}. This results into reduction of the switch resistance (R_s) and increase in I_{DD}. When the monitored I_{DD} is larger than the reference, the switch control voltages are not changed. This unipolar compensation techniques tracks the I_{DD} degradation and tries to compensate it.

However, the threshold voltage shift due to the aging degradation show quasi saturation behavior due to power law relationship towards stress time (t^n), see equations (2.1) and (2.2). Hence the degradation is higher at the beginning of the stress and then it partially saturates after long stress time. Based on this knowledge the unipolar compensation technique is modified to adaptive bipolar compensation technique. In adaptive bipolar technique at the beginning of the accelerated stress condition ($t < 100s$), where more degradation is expected, large V_{step} ($V_{step_L} = 100\mu V$) is used. For stress time ($100s < t < 1000s$) medium V_{step} ($V_{step_M} = 50\mu V$) is used. And for later stress time ($t > 1000s$), where less degradation is expected, small V_{step} ($V_{step_S} = 25\mu V$) is used. In addition, when the monitored I_{DD} is larger than the reference, depending on the stress time, $\frac{V_{step_{L,M,S}}}{2}$ is subtracted from V_{NEN} and added to V_{PEN} in order to reduce the time required to track the reference. Figure 6.8 illustrates the adaptive bipolar compensation technique algorithm used to monitor and compensate aging induced performance degradation.

6.3 Measurements

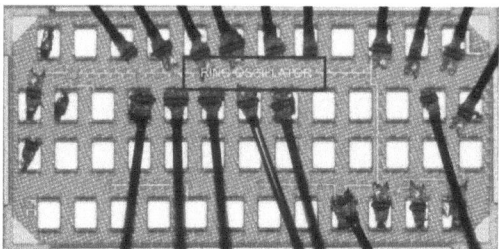

Fig. 6.9: Die photograph of ring oscillator test chip

To prove this basic concept of the background monitoring and compensation of aging induced performance degradation, measurements are performed on the ring oscillator test chips. These test chips are fabricated using 32nm high-κ, metal gate CMOS technology. The die photograph of the measured test chip is illustrated in figure 6.9. Accelerated aging conditions similar to that used during simulations, i.e. stress time of 10^4s at $V_{DD} = 1.481V$ and $T = 125°C$, are applied to the samples. The switch transistors control voltage ($|V_{gs}|_{n/p}$) was set to 0.9V at the beginning of compensation to avoid its aging degradation. The monitoring of performance degradation and compensation during aging was achieved using a completely automated measurement test setup illustrated in figure 6.10, controlled using a LabVIEW program.

The quasi-saturation behavior of aging degradation was observed in figure 6.11, where I_{DD} degradation in ring oscillator circuit without and with compensation is measured during stress. With adaptive bipolar compensation the degradation in I_{DD} is significantly reduced. Figure 6.12 illustrates the comparison between tracking accuracy of unipolar and adaptive bipolar compensation technique. Unipolar technique has two main drawbacks. The compensation is found to be lagging in the beginning of the stress period due to larger degradation compared to the used voltage step ($V_{step} = 50\mu V$) values. And during the later stress period it leads or over compensates the reference since the same V_{step} are now larger compared to the degradation values. The adaptive bipolar compensation technique

Measurements

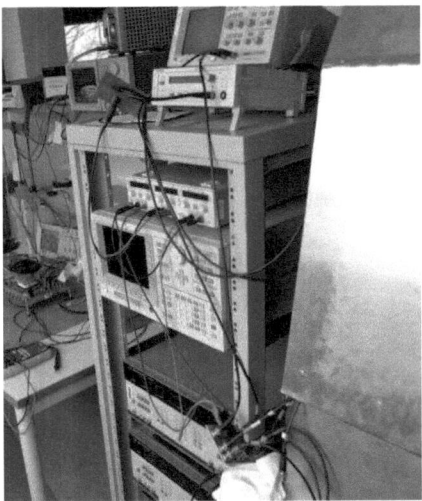

Fig. 6.10: Measurement test setup for accelerated aging of ring oscillator test chip

efficiently takes care of both these drawbacks. Using adaptive bipolar compensation technique the aging degradation is compensated by 98% over the accelerated lifetime of

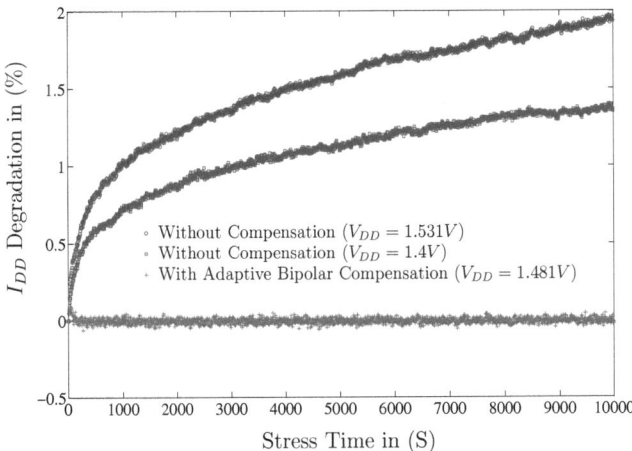

Fig. 6.11: Degradation of the ring oscillator supply current (I_{DD}) in percentage over stress time without and with using adaptive bipolar compensation technique, characterized at stress voltage and temperature

68   Chapter 6. Aging in Ring Oscillator Circuits

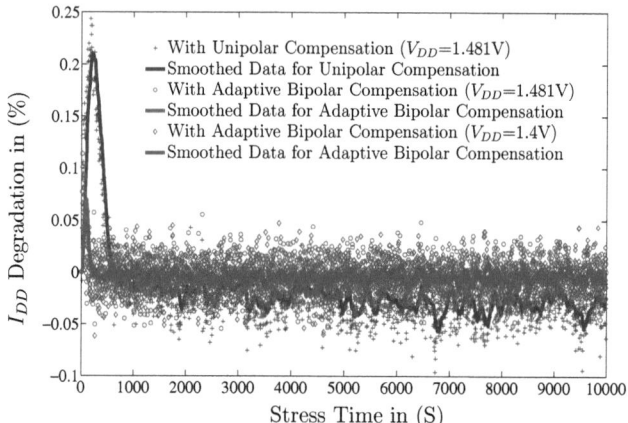

Fig. 6.12: Improvement in compensation of ring oscillator supply current (I_{DD}) degradation over stress time using adaptive bipolar compensation technique, characterized at stress voltage and temperature

this ring oscillator circuit. Moreover, the compensation is achieved within 200s using adaptive bipolar tracking compared to 600s required for unipolar tracking. Figure 6.13 illustrates the comparison between switch control voltage tracking behavior during the stress period under unipolar and adaptive bipolar compensation techniques. It can be

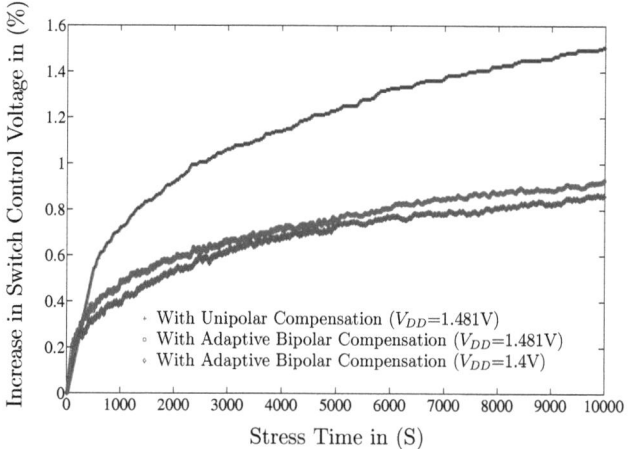

Fig. 6.13: Switch control voltage ($|V_{gs}|_{n/p}$) tracking during stress, characterized at stress voltage and temperature

observed that unipolar technique results into over compensation. This is because except at the beginning of the stress period, the compensation technique leads the reference which results into higher f_{osc} and hence results into higher degradation.

The device degradation due to stress consists of so-called permanent and relaxation components as discussed in section 2.2.1. The permanent component leads to slow varying V_{th} and I_d shifts on device level. The relaxation component is caused by de-trapping of charge during subsequent recovery phase after stress. This recovery process has a broad range of time constants. The adaptive bipolar tracking algorithm discussed in this chapter can be implemented on-chip to monitor not only on-the-fly aging degradation behavior but also recovery behavior. To measure the recovery behavior stress phase (without compensation) can be performed, followed by the fast tracking of relaxation behavior during recovery phase using the switch control voltages. The change in switch control voltages will provide precise information on the recovery behavior in the time range that depends on how fast the performance can be monitored and the compensation steps can be applied.

6.4 Summary

In this chapter the contribution of different aging mechanisms towards parameter drifts in transistors of the ring oscillator circuit implemented in 32nm high-κ metal gate CMOS technology, stressed under both AC and DC stress was presented. It was shown by aging simulation and measurement results that unlike in previous technology nodes, enhanced HCI mechanism under AC stress leads to higher performance degradation in the ring oscillator circuit compared to BTI degradation and hence significant lifetime enhancement under AC stress is no more guaranteed. Also it was demonstrated by measurement results that during characterization the relative performance degradation of the ring oscillator circuit increases with decreasing operating voltage (V_{DD}) due to reduced transistor voltage headroom and overdrive.

A very simple but effective technique to monitor and compensate aging degradation in ring oscillator circuit was presented. An improved adaptive bipolar algorithm to achieve quick and accurate compensation of aging degradation was suggested. The step size for compensation was adapted based on the quasi-saturation behavior of aging degradation under adaptive bipolar technique. This concept was demonstrated using a modified ring oscillator circuit test chip. The aging degradation was compensated by around 98% over the lifetime of this ring oscillator circuit. This tracking algorithm could further be implemented on-chip to monitor and compensate aging degradation on-the-fly. Also recovery behavior could be monitored with this concept.

Chapter 7

Aging in Switches used in Switched Capacitor Circuits

This chapter provides evaluation of aging degradation in switches used in the switched capacitor (SC) circuits. Switched capacitor circuits are sampled data or discrete time circuits most commonly used in implementing filters, analog to digital and digital to analog converters. Aging leads to increase in switch resistance over lifetime resulting into limited transfer of charge onto capacitor at high operating frequencies. The aging degradation in switches is studied using a special ring oscillator test chip with switch stages added at the input of each inverter stage. The aging of the switch results into degradation of the ring oscillator and hence frequency reduction. The amount of degradation and its impact on circuit performance is investigated in this chapter.

7.1 Switches in SC Circuit

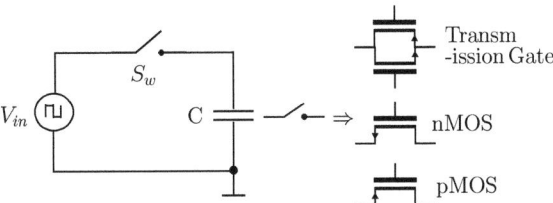

Fig. 7.1: Schematic of basic switched capacitor circuit and illustration of different types of switches

A simple switched capacitor circuit example is a sampling circuit illustrated in figure 7.1, which is the most basic building block of a sampled data analog to digital converter [73]. A basic sampling circuit consists of a switch (S_w) and a capacitor (C). The switch can be implemented using nMOSFET, pMOSFET or complementary (both

Chapter 7. Aging in Switches used in Switched Capacitor Circuits

transistors) MOSFET devices. The switch is controlled (either ON or OFF) using gate control voltage of V_{DD} and V_{SS} depending on the type of the device used. The ON resistance (R_{on}) of the nMOSFET transistor increases considerably as input voltage approaches above ($V_{DD}-V_{thN}$) and for pMOSFET transistor when input voltage approaches below $|V_{thP}|$. Hence most commonly complementary or transmission (Tr) gate topology is used for switch implementation to enable rail-to-tail voltage swings. The R_{on} of a Tr gate switch is given by equation (7.1). Ignoring body effect, the R_{on} of Tr gate is ideally independent of the input voltage level in the first order of approximation for high V_{DD} values.

$$R_{on} = R_{onN} \parallel R_{onP}$$

$$R_{on} = \frac{1}{\mu \cdot C_{ox} \cdot (\frac{W}{L}) \cdot (V_{DD} - V_{thN} - |V_{thP}|)} \quad (7.1)$$

when $\mu_N \cdot C_{ox} \cdot (\frac{W}{L})_N = \mu_P \cdot C_{ox} \cdot (\frac{W}{L})_P$
where μ is the charge-carrier effective mobility, W is the transistor gate width, L is the transistor gate length and C_{ox} is the gate oxide capacitance per unit area.

The sampling speed of the switch capacitor circuit depends on the value of R_{on} and C. The voltages across the capacitor (V_C) and switch (V_S) in figure 7.1 when capacitor is charging from 0V, are given by equations (7.2) and (7.3) respectively. For discharging the equations are interchanged.

$$V_C(t) = V_{in} \cdot (1 - e^{\frac{-t}{\tau}}) \quad (7.2)$$

$$V_S(t) = V_{in} \cdot e^{\frac{-t}{\tau}} \quad (7.3)$$

$$\tau = R_{on} \cdot C \quad (7.4)$$

where τ is the time required by V_C to reach $V_{in} \cdot (1 - \frac{1}{e})$, i.e. C to charge to about 63.2%. Hence about 5τ are required for C to be fully charged (99.3%).

The aging related performance degradation of the Tr gate switch used in the switched capacitor circuit is evaluated for a mobile phone use case of 4 years at 85°C with 105% of maximum specified V_{DD} ($V_{DD} = 1.155V$). The aging simulations were performed both under static (DC) stress and dynamic (AC) stress at frequency of 1MHz with 50%

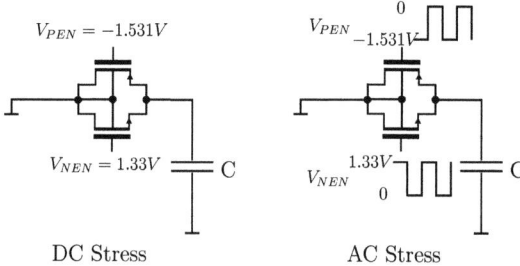

DC Stress   AC Stress

Fig. 7.2: **Accelerated aging DC and AC stress conditions for Tr gate switch used in the switched capacitor circuit**

Switches in SC Circuit 73

duty cycle. In order to continuously degrade both the transistors in the switch, for emulating the worst case stress condition, $V_{in} = 0V$ and $V_{DD} = 0V$ were used during aging simulation, as illustrated in figure 7.2. Hence, the equivalent accelerated aging bias condition for stress time of 10^4s at 125°C was derived to be $V_{NEN} = 1.33V$ and $V_{PEN} = -1.531V$.

Figure 7.3 shows the contribution of different aging mechanisms in nMOSFET and pMOSFET transistors of the Tr gate switch under DC and AC use case and accelerated stress condition. Both the transistors are affected only by BTI aging mechanism. The BTI degradation in pMOSFET transistor is dominant compared to that of nMOSFET transistor. And it is reduced by around 50% under AC stress compared to that under DC stress considering the AC factor due to the recovery effect. CHCI and NCHCI degradations are absent due to zero input voltage. Moreover the Tr gate switch operates in the triode region for most of the time hence CHCI and NCHCI degradations are negligible even for non-zero input voltages.

$$\Delta R_{on}\% = \frac{R_{on(post-stress)} - R_{on(pre-stress)}}{R_{on(pre-stress)}} \cdot 100 \qquad (7.5)$$

Figure 7.4 illustrates the comparison between simulated R_{on} of Tr gate switches over input voltage range before and after aging under AC and DC stress conditions, characterized at nMOSFET switch control voltage ($V_{NEN} = 0.9V$), pMOSFET switch control voltage ($V_{PEN} = 0V$), $V_{DD} = 0.9V$ and $T = 25$°C. The relative degradation in the aged switch (ΔR_{on}) increases as the switch overdrive reduces during characterization due to reduced transistor voltage headroom. This behavior is presented in table 7.1 where relative degradation of R_{on} is evaluated using equation (7.5) characterized at different V_{in} values. Here maximum degradation is observed when $V_{in} \approx 0.5V$ i.e. the overdrive of the

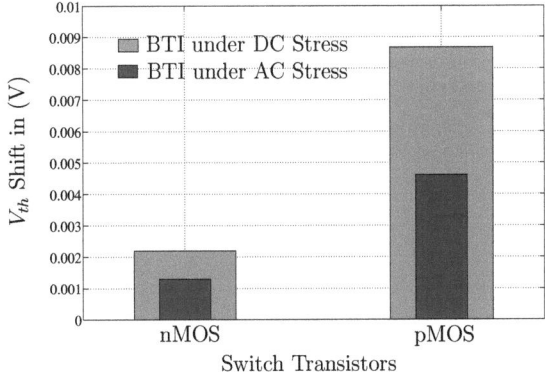

Fig. 7.3: Contribution of different aging mechanisms causing degradation in transmission gate switch under both DC and AC stress for mobile phone EoL use case and accelerated condition

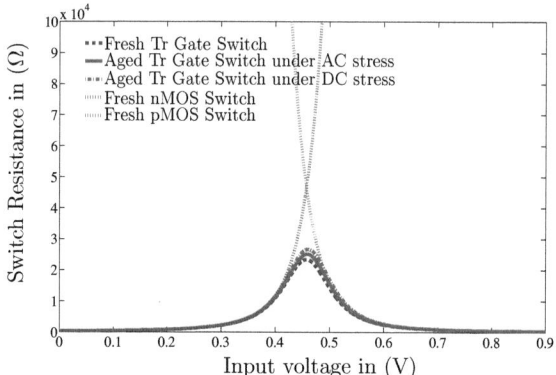

Fig. 7.4: Comparison between simulated "ON" resistance (R_{on}) of the fresh and aged (under DC and AC stress for mobile phone EoL use case condition) transmission gate switches, characterized at different input voltage values for $V_{NEN} = 0.9V$, $V_{PEN} = 0V$, $V_{DD} = 0.9V$ and $T = 25°C$

devices in the switch is close to zero ($V_{th} \approx 0.4V$). Figure 7.5 illustrates the voltage across the capacitor resulting from charging and discharging cycles for a pulsed and rising input signal, both before and after aging. A realistic behavior of switched capacitor circuit used in ADC's where the capacitors are charged and discharged with input signal with respect to analog ground (0.5V) is illustrated in figure 7.5(b). Figure 7.4 and 7.5 illustrate that aging leads to increase in R_{on} of the Tr gate switch and hence increase in charging and discharging time constant (τ) which leads to limited transfer of charge onto capacitor. The circuit operation is severely affected when the aged switches operate at low overdrive voltages.

7.2 Aging Monitor for Switch Degradation

To monitor the degradation in the switches under various aging mechanisms, the same ring oscillator test structure described in section 6.2 and illustrated in figure 7.6, is

Characterized at → Stress Condition ↓	$V_{in} = 0V$	$V_{in} = 0.4V$	$V_{in} = 0.5V$	$V_{in} = 0.9V$
Use case (DC)	0.537	5.34	19.2	1.76
Use case (AC)	0.322	3.07	9.63	0.92

Table 7.1: Simulated relative increase in R_{on} (ΔR_{on}) in percentage for the transmission gate switches aged under DC and AC stress for mobile phone EoL use case condition, evaluated at different input voltage values for $V_{NEN} = 0.9V$, $V_{PEN} = 0V$, $V_{DD} = 0.9V$ and $T = 25°C$

Aging Monitor for Switch Degradation

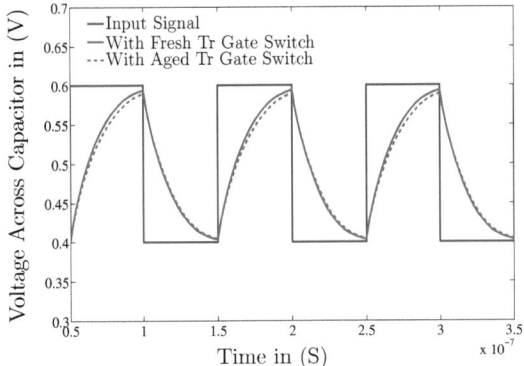

(a) Pulsed Input Signal at 10MHz, 50% duty cycle and switch turned ON continuously (C=2pf)

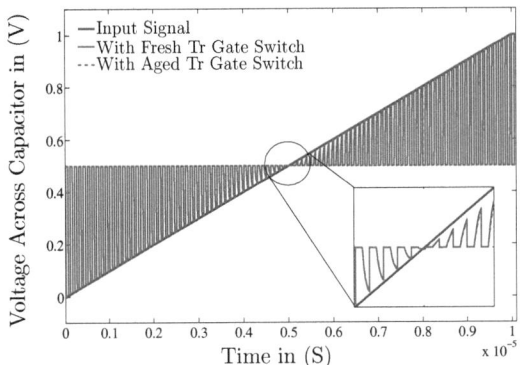

(b) Rising input signal and switch turned ON and OFF alternately at 10MHz, 50% duty cycle and C=2pf

Fig. 7.5: Transient waveform of voltage across capacitor during charging and discharging cycles with fresh and aged (under DC stress for mobile phone EoL use case condition) transmission gate switch

used for aging simulation and measurements. Here each inverter circuit stage models the capacitor and is not aged by applying $V_{DD} = 0V$. Hence the degradation in the ring oscillator frequency (f_{osc}) post aging directly represents degradation in the switches. The detailed schematics of two inverter and switch stages with the (W/L) ratios of the MOSFET devices are shown in figure 7.7 and table 7.2 respectively. There are total 23 inverter and switch stages (N) in the circuit. The frequency divider circuit is added to divide the f_{osc} by a frequency divider ratio ($Divider$) of 2^{10} which makes the measurement

Chapter 7. Aging in Switches used in Switched Capacitor Circuits

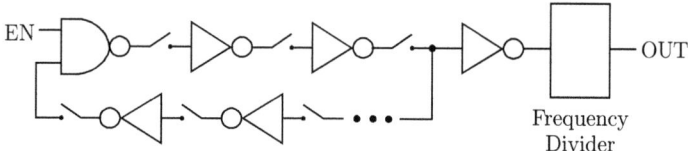

Fig. 7.6: Schematic of test structure used for CMOS switch aging measurement

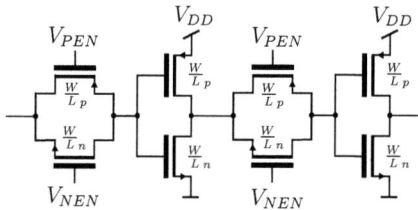

Fig. 7.7: Schematic of two stages of CMOS switch and inverter

Transistor	W/L [nm/nm]
pMOS ($\frac{W}{L}_p$)	1020/30
nMOS ($\frac{W}{L}_n$)	690/30

Table 7.2: MOSFET devices W/L ratios

setup simple. The frequency of this ring oscillator circuit is given by equation (7.6):

$$f_{osc} = \frac{1}{2 \cdot N \cdot \tau_d \cdot Divider}$$

$$\tau_d = \frac{\tau_n + \tau_p}{2} + \tau_s \qquad (7.6)$$

$$\tau_s = R_s \cdot C_L$$

where, τ_n, τ_p are the switching times of nMOSFET and pMOSFET transistors and R_s is the equivalent resistance of the switch stage. Aging leads to increase in R_s and hence degradation of f_{osc}.

7.3 Measurements

To evaluate the degradation of the transmission gate switches measurements are performed on the ring oscillator test chips fabricated in 32nm high-κ metal gate CMOS technology. The die photograph of the measured test chip and the measurement setup is illustrated in figures 6.9 and 6.10 respectively. Accelerated aging conditions similar to that used during simulations, i.e. stress time of 10^4s at $V_{NEN} = 1.33V$, $V_{PEN} = -1.531V$

and $T=125°C$, are applied to the samples. The supply voltage (V_{DD}) is set to 0V to avoid the ring oscillator aging degradation. Two test chips are stressed under DC and AC (1MHz with 50% duty cycle) accelerated stress conditions respectively. The entire measurement activity is divided into following steps and is repeated for both samples:

$$M1 \Rightarrow M2 \Rightarrow S \Rightarrow R \Rightarrow M3 \Rightarrow M4 \Rightarrow A \Rightarrow M5 \Rightarrow M6$$

where,
"S" → Stress phase with stress time of 10^4s at $125°C$ and $V_{NEN} = 1.33V$, $V_{PEN} = -1.531V$, $V_{DD} = 0V$
"R" → Relaxation phase with stress time of 10^3s at $125°C$ and $V_{NEN} = 0.7V$, $V_{PEN} = 0.2$, $V_{DD} = 0.9V$
"A" → Annealing phase at high temperature, $125°C$ and $V_{NEN} = 0V$, $V_{PEN} = 0V$, $V_{DD} = 0V$ for 10^4s to study long term permanent offset behavior

And the f_{osc} and I_{DD} measurements were performed,
"M1" → before stress at $25°C$ and $V_{DD} = 0.9V$
"M2" → before stress at $125°C$ and $V_{DD} = 0.9V$
"M3" → post stress at $125°C$ and $V_{DD} = 0.9V$
"M4" → post stress at $25°C$ and $V_{DD} = 0.9V$
"M5" → post annealing at $125°C$ and $V_{DD} = 0.9V$
"M6" → post annealing at $25°C$ and $V_{DD} = 0.9V$.

Figure 7.8 illustrates the behavior of drain current (I_{DD}) degradation in the ring oscillator structure due to aging of switches under DC and AC stress conditions measured during different measurement phases evaluated at $125°C$ for $V_{DD} = 0.9V$, $V_{NEN} = 0.7V$ and $V_{PEN} = 0.2V$. A degradation of 2.1% under DC stress and 1.132% under AC stress are noted after stress phase and before relaxation phase. It can be confirmed with these

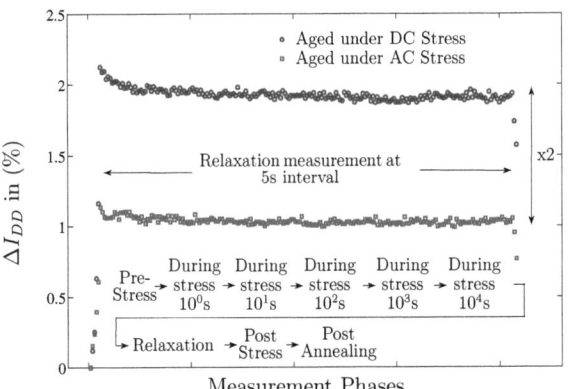

Fig. 7.8: Measured I_{DD} degradation due to aging of switches under DC and AC stress $(V_{DD} = 0V)$ characterized at $125°C$ for $V_{DD} = 0.9V$, $V_{NEN} = 0.7V$, $V_{PEN} = 0.2V$ for different measurement phases

78   Chapter 7. Aging in Switches used in Switched Capacitor Circuits

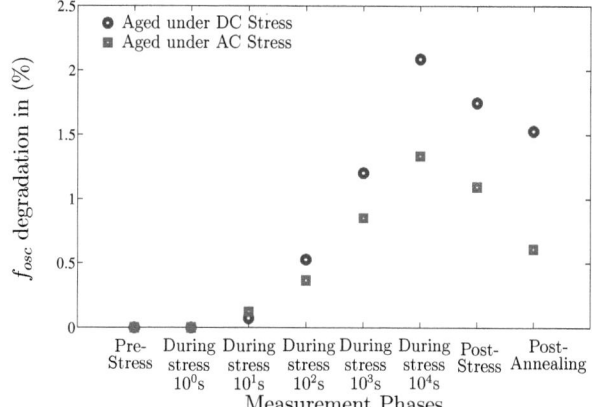

Fig. 7.9: Measured f_{osc} degradation due to aging of switches under DC and AC stress ($V_{DD} = 0V$) characterized at 125°C for $V_{DD} = 0.9V$, $V_{NEN} = 0.7V$, $V_{PEN} = 0.2V$ for different measurement phases

measurement results that BTI degradation under DC stress is ≈2x times more compared to that under AC stress. Relaxation in degradation is observed after stress due to BTI recovery effect. These measurements during the relaxation phase are performed at 5s interval. A relaxation of 0.563% after DC stress and 0.37% after AC stress are measured in the post annealing phase.

Figure 7.9 illustrates the behavior of ring oscillator frequency (f_{osc}) degradation

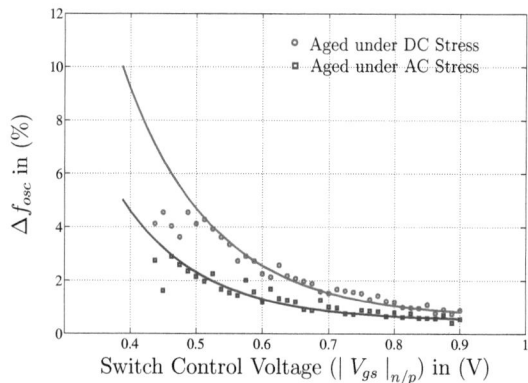

Fig. 7.10: Measured relative degradation of the ring oscillator frequency (Δf_{osc}) in percentage vs. switch control voltage ($|V_{gs}|_{n/p}$) characterized at $V_{DD} = 0.9V$ and $T = 125°C$

due to aging of switches under DC and AC stress conditions measured during different measurement phases evaluated at 125°C for $V_{DD} = 0.9V$, $V_{NEN} = 0.7V$ and $V_{PEN} = 0.2V$. Degradation values similar to I_{DD} degradation are noted post stress and annealing. A degradation of 2.089% under DC stress and 1.333% under AC stress are noted after stress phase and before relaxation phase. It can be confirmed from the measurement results plotted in figures 7.8 and 7.9 that aging leads to increase in switch resistance that results into f_{osc} and I_{DD} degradation.

$$\Delta f_{osc}\% = \frac{f_{osc(pre-stress)} - f_{osc(post-stress)}}{f_{osc(pre-stress)}} \cdot 100 \qquad (7.7)$$

Further, during characterization the relative performance degradation of the aged switched capacitor circuit increases as the switch overdrive voltage decreases due to reduced transistor voltage headroom. This is depicted in figure 7.10, where relative degradation f_{osc} is evaluated using equation (7.7) characterized at different switch control voltage values. It can be observed that the relative degradation increases as switch control voltage reduces. However, one of the transistors in the Tr gate switch turns "OFF" below $|V_{gs}|_{n/p} = 0.5V$ after which this relation does not hold true.

Hence it can be concluded that the performance degradation of the switched capacitor circuit due to aging of switch depends not only on the magnitude of switch degradation but also on the circuit operating frequency and overdrive of the switch during circuit operation.

7.4 Countermeasures

Recovery of BTI degradation in both nMOSFET and pMOSFET devices under accumulation stress was reported in [46]. Recovery in case of nMOSFET device whereas degradation in case of pMOSFET device was measured under accumulation stress by [5, 74]. Hence a possibility of using bipolar AC stress during aging as a countermeasure to compensate or reduce degradation is considered. With bipolar stress both the transistors in the switch experience inversion stress during half clock period and accumulation stress during the remaining half period. The degradation induced by inversion stress is expected to be fully or partially compensated by accumulation stress.

The behavior of BTI degradation under accumulation stress is not modeled in the sub-circuit aging simulation models discussed in section 2.3. However in [75] behavior of BTI degradation under DC, unipolar and bipolar stress was investigated. It was shown using measurements that for bipolar stress additional interface state trap generation can result into enhanced BTI degradation. This effect is enhanced at high frequencies. To confirm this behavior measurements are performed on the switches in the ring oscillator test chip under bipolar AC (1MHz, 50% duty cycle) stress for stress time of 10^4s at 125°C, $V_{DD} = 0V$, $V_{NENhigh} = 1.5V$ $V_{NENlow} = -1.5V$ and $V_{PEN} = 0V$, such that only nMOSFET was aged in both inversion and accumulation region. As illustrated in figure 7.11, no significant reduction in f_{osc} degradation is observed under bipolar stress

80   Chapter 7. Aging in Switches used in Switched Capacitor Circuits

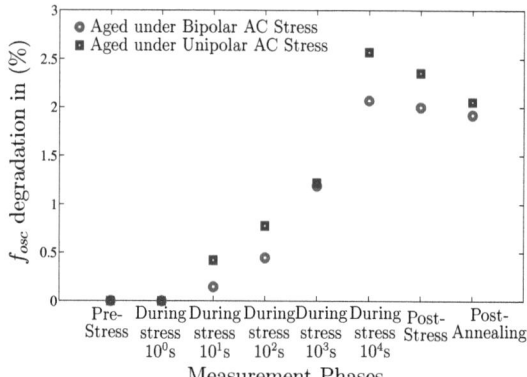

Fig. 7.11: Measured f_{osc} degradation due to aging of nMOSFET device in transmission gate switch under bipolar and unipolar AC stress ($V_{DD} = 0V$) characterized at 125°C for $V_{DD} = 0.9V$, $V_{NEN} = 0.5V$, $V_{PEN} = 0.4V$ for different measurement phases

as compared to unipolar stress.

Hence to ensure proper working of switched capacitor circuit under aging degradation of switches, care must be taken to appropriately size the transistors during design phase based on the aging information to always remain below the maximum RC time constant value required for operating clock frequency range.

7.5 Summary

In this chapter the contribution of different aging mechanisms towards parameter drifts in transistors of the transmission gate switches used commonly in switched capacitor circuits was presented. The test structure was implemented in 32nm high-κ metal gate CMOS technology and stressed under both AC and DC stress. It was found by aging simulation and measurements that both the transistors in the switch are affected by BTI aging mechanism and it is reduced by around 50% under AC stress compared to that under DC stress due to recovery effect.

The relative performance degradation of the switch increases as the switch overdrive voltages during characterization is decreased, due to reduced voltage headroom. It was shown that the performance degradation of the switched capacitor circuit due to aging depends not only on the magnitude of switch degradation but also on the circuit operating frequency and overdrive of the switch during circuit operation. The ineffectiveness of using bipolar stress as on-line countermeasure to compensate aging degradation using accumulation stress was discussed. And hence the importance of proper sizing of switches to ensure reliable working of switched capacitor circuit under aging degradation was highlighted.

Chapter 8

Aging in Successive Approximation Register ADC

Successive approximation register (SAR) analog to digital converter (ADC) is a widely used Nyquist rate ADC in the field of medium to high resolution applications. These ADC's can be designed for high-performance at low-power [76]. The SAR ADC works on the principle of binary search algorithm which is most commonly implemented using charge redistribution technique. A typical high performance SAR ADC consist of an input buffer circuit to drive input and reference voltages, comparator, capacitive digital to analog converter (DAC) and SAR control logic. In this chapter the effect of aging degradation mechanisms on performances of a state-of-the-art fully differential 12-bit SAR ADC are discussed.

8.1 Introduction to SAR ADC

The performance of SAR ADC is very sensitive to non-idealities in the input buffer, DAC and comparator circuits. Aging mechanisms like NBTI, PBTI, CHCI and NCHCI induce mismatch in matched transistor pairs resulting into circuit performance degradation. A detailed analysis and discussion regarding effect of aging induced transistor parameter drifts on the performances of closed and open loop OTA circuits implemented using 32nm high-κ metal gate CMOS technology are presented in chapter 4. Hence based on these findings, the aging degradation of SAR ADC performance is analyzed in this chapter.

The schematic of 12-bit fully differential SAR ADC is illustrated in figure 8.1 [63, 77]. It consists of a fully differential input buffer circuit, two binary weighted capacitor arrays, a fully differential comparator and a SAR control logic module. The input buffer circuit is used in the charge redistribution based high speed SAR ADC to drive the DAC within a short settling time ($\approx$4ns). The buffer circuit is implemented using an OTA circuit in closed loop configuration. The DAC is charge redistribution based and hence is implemented using capacitors. The buffer drives the DAC with reference voltage ($Vref_{p/n}$) and analog input ($Ain_{p/n}$) during different phases of the clock signals that

Chapter 8. Aging in Successive Approximation Register ADC

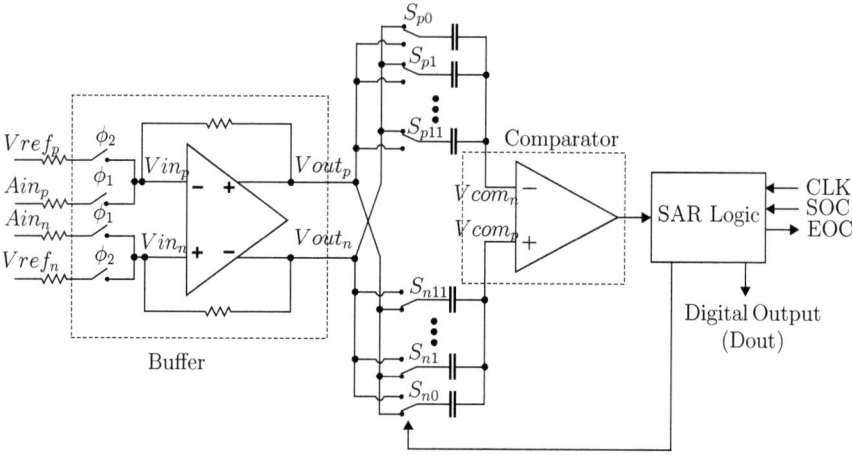

Fig. 8.1: Schematic of Successive Approximation Register ADC

control the switches. The clock signals are shown only schematically in figure 8.2, not displaying the more complex clocking during the ADC operation. The comparator used in the SAR ADC typically consist of few preamplifier stages realized using open loop OTA, followed by a regenerative latch stage which uses positive feedback. This first open loop OTA stage of the comparator mainly determines the precision of the overall comparator.

The operation of the circuit can be divided into three phases. The first phase is offset cancellation and sampling phase. The second phase is the hold phase. And the third phase is the bit-cycling phase. At the beginning of each conversion start of conversion (SOC) signal indicates the SAR logic to begin the conversion process and at the end of conversion the end of conversion (EOC) signal indicates the user that the digital output of converted analog input signal is available.

During the offset cancellation and sampling phase, the preamplifier stages in the comparator circuit are configured in closed loop configuration. So the two inputs of the comparator are at same potential equal to the common mode voltage ($Vcm = 0.5V$), which is required for offset cancellation. The input of the buffer circuit is connected to Ain_p and Ain_n, hence signal at Ain_p is connected to $Vout_p$ and at Ain_n is connected to $Vout_n$. The switches S_{px} (x=0 to 11) are connected to $Vout_p$ and the switches S_{nx} (x=0 to 11) are connected to $Vout_n$. Hence the two capacitor arrays are charged respectively to voltages of ($\pm \frac{Ain_p - Ain_n}{2}$).

During the hold phase, the preamplifier stages in the comparator circuit are configured in open loop configuration. And then the input of the buffer circuit is connected to $Vref_p$ and $Vref_n$, hence signal at $Vref_p$ is connected to $Vout_p$ and at $Vref_n$ is connected to $Vout_n$. Also the switches S_{px} (x=0 to 11) are connected to $Vout_n$ and the switches S_{nx} (x=0 to 11) are connected to $Vout_p$. Hence the voltage at the negative input terminal of

Introduction to SAR ADC

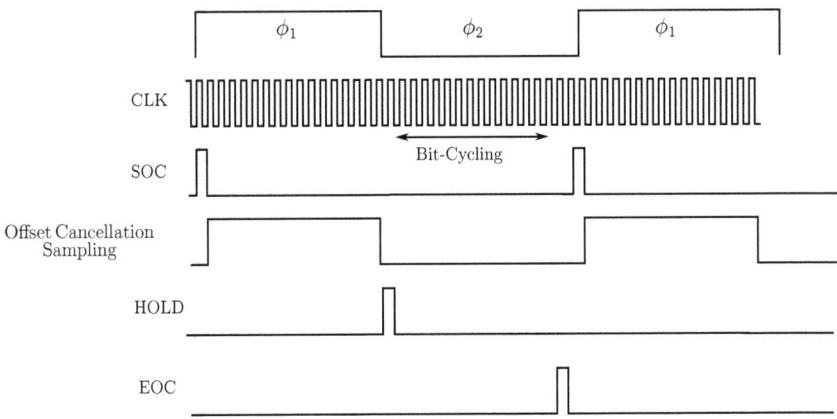

Fig. 8.2: SAR ADC control signals

the comparator $(Vcom_n)$ is $(Vref_n - \frac{Ain_p - Ain_n}{2})$. While the voltage at the positive input terminal of the comparator $(Vcom_p)$ is $(Vref_p + \frac{Ain_p - Ain_n}{2})$.

During the bit-cycling phase, the SAR logic performs bit-cycling for 12 clock cycles to perform the successive approximation. The input of the buffer circuit remains connected to $Vref_p$ and $Vref_n$, hence $Vout_p$ is connected to $Vref_p$ and $Vout_n$ is connected to $Vref_n$. At the beginning switch S_{p11} is connected from $Vout_n$ to $Vout_p$. Hence the voltage at $(Vcom_n)$ is added with $(\frac{vref_p - vref_n}{2})$. And the switch S_{n11} is connected from $Vout_p$ to $Vout_n$. Hence the voltage at $(Vcom_p)$ is subtracted with $(\frac{vref_p - vref_n}{2})$. The other switches remain unchanged. The output of the comparison of these two input voltages decides the MSB (12th bit). If the comparison result is high, switch S_{p11} remains connected to $Vout_p$ and switch S_{n11} remains connected to $Vout_n$ throughout the conversion. On the other hand if the evaluation result is low, switch S_{p11} is connected back to $Vout_n$ and switch S_{n11} is connected back to $Vout_p$. This completes evaluation of the MSB bit. The evaluation of the remaining bits is done similar to the MSB evaluation.

8.1.1 SAR ADC Model Implementation

To study the impact of aging degradation in different building blocks on the performance of a SAR ADC circuit, all the blocks viz. buffer, DAC, comparator and SAR control logic, of the 12-bit SAR ADC as illustrated in figure 8.1 are modeled as ideal elements using Verilog-A to reduce the complexity and the simulation time. Based on the working of SAR ADC as explained in section 8.1 the model implementation is done as follows:

The voltages at different nodes during offset cancellation and sampling phase are

Chapter 8. Aging in Successive Approximation Register ADC

given by equation 8.1:

$$Vcom_p = Vcm$$
$$Vcom_n = Vcm$$
$$dac_p = \frac{Ain_p - Ain_n}{2}$$
$$dac_n = \frac{Ain_n - Ain_p}{2} \tag{8.1}$$

where dac_p and dac_n are the voltages across the two capacitor arrays.

The voltage at different nodes during the hold phase is given by equation 8.2:

$$Vcom_p(1) = Vref_n - dac_p$$
$$Vcom_n(1) = Vref_p - dac_n \tag{8.2}$$

Finally the voltage at different nodes during the bit-cycling phase is given by equation 8.3:

$$Vcom_p(2) = Vcom_p(1) + \frac{Vref_p - Vref_n}{2}$$
$$Vcom_n(2) = Vcom_n(1) - \frac{Vref_p - Vref_n}{2}$$
$$for\ i = 2 : 1 : 13$$
$$\quad if\ (Vcom_p(i) < Vcom_n(i))$$
$$\qquad Vcom_p(i+1) = Vcom_p(i) + \frac{Vref_p - Vref_n}{2^i}$$
$$\qquad Vcom_n(i+1) = Vcom_n(i) - \frac{Vref_p - Vref_n}{2^i}$$
$$\qquad Dout(14 - i) = 1$$
$$\quad else$$
$$\qquad Vcom_p(i+1) = Vcom_p(i) - \frac{Vref_p - Vref_n}{2^i}$$
$$\qquad Vcom_n(i+1) = Vcom_n(i) + \frac{Vref_p - Vref_n}{2^i}$$
$$\qquad Dout(14 - i) = 0 \tag{8.3}$$

where $Dout$ is the final digital output in binary format.

The converter is simulated with a clock frequency of 33.33MHz, and a conversion rate of 1.19MS/s was used with $V_{refp} = 0.8V$, $V_{refn} = 0.2V$, $V_{in} = 0.6V_{p-p}$ and $V_{DD} = 1V$.

8.2 Aging in SAR ADC Building Blocks

The aging induced input referred offsets in closed and open loop simple Miller OTA circuit configuration analyzed in chapter 4 are summarized in table 8.1 and 8.2 respectively.

Effect of Aging on SAR ADC Performance

Stress Time	Temp (°C)	V_{scale} (%)	Offset (mV)
4 Yrs	85	105% (V_{DD} = 1.155V)	0.012
10 Yrs	125	105% (V_{DD} = 1.155V)	0.0305

Table 8.1: Simulated aging induced input referred offset resulting from asymmetrical DC input stress ($Vin_n = 1.155V$, $Vin_p = 0V$) in closed loop simple Miller OTA configuration evaluated at $V_{DD} = 1V$ and $T = 25°C$

Stress Time	Temp (°C)	V_{scale} (%)	Offset (mV)
4 Yrs	85	105% (V_{DD}=1.155V)	5.29
10 Yrs	125	105% (V_{DD}=1.155V)	10.438

Table 8.2: Simulated aging induced input referred offset resulting from asymmetrical DC input stress ($Vin_n = 1.155V$, $Vin_p = 0V$) in open loop simple Miller OTA configuration evaluated at $V_{DD} = 1V$ and $T = 25°C$

The transistor parameter shifts due to aging degradation mechanisms induces mismatch in matched differential pairs in the closed and open loop OTA circuit configurations. The most degraded circuit performance is offset. In the input buffer circuit used in the SAR ADC, the OTA always operates in closed loop configuration and its input transistors see smaller stress compared to the output stage transistors. Simulation results show that these output transistors are the main contributors to offset. On the other hand in the preamplifier stages in the comparator used in the SAR ADC, the OTA operates in open loop configuration and its input transistors see a large stress. Simulation results show that these input transistors are the main contributors to offset. The effect of these aging induced offsets in the buffer and comparator circuit on performance of SAR ADC is incorporated in the models and analyzed in the next subsection. The implemented DAC is charge redistribution based and since the capacitors are not affected by HCI and BTI wearout mechanisms it is considered ideal in this analysis. The impact of switch aging is neglected since proper device sizing can take care of the aging induced increase in switch "ON" resistance as discussed in chapter 7.

8.3 Effect of Aging on SAR ADC Performance

The aging induced performance degradation in the closed and open loop OTA configurations are incorporated into the 12-bit SAR ADC model which is introduced in section 8.1.1. This section presents SAR ADC performance degradation due to aging of its building blocks particularly the input buffer and the comparator circuits. The individual and combined effects are discussed separately.

8.3.1 Effect of Buffer Aging on SAR ADC

Aging of the buffer circuit under asymmetrical input stress for mobile phone EoL use case condition induces input referred offset of $12\mu V$, as presented in table 8.1. This offset value

86   Chapter 8. Aging in Successive Approximation Register ADC

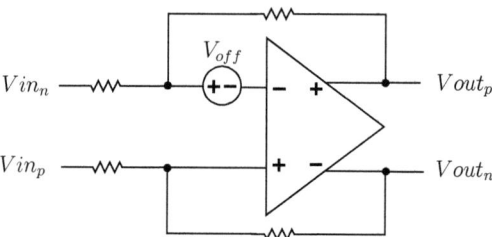

Fig. 8.3: Schematic of input buffer circuit with modeled aging induced offset

is very small when compared to the resolution (V_{LSB}) of a 12-bit SAR ADC, evaluated using equation (8.4) which is around 146.5μV for full scale voltage (V_{FS}) range of 0.6V_{p-p}. Hence this offset does not affect the ADC performance as long as it remains below $\frac{1}{2}V_{LSB}$. In case the offset becomes larger than this value, then the transfer characteristics depicted in figure 8.4 shows that a gain error arises in the converter due to aging induced offset in the input buffer circuit. The deviation from ideal curve is maximum at small values of input signal and reduces gradually until the input reaches its full-scale value. This results from the fact that the input signals ($Ain_{p/n}$) see complete offset voltage of the input buffer during offset cancellation and sampling phase, whereas the reference voltage ($Vref_{p/n}$) see offset voltage which is divided by 2^i during each successive approximation cycle of bit-cycling phase.

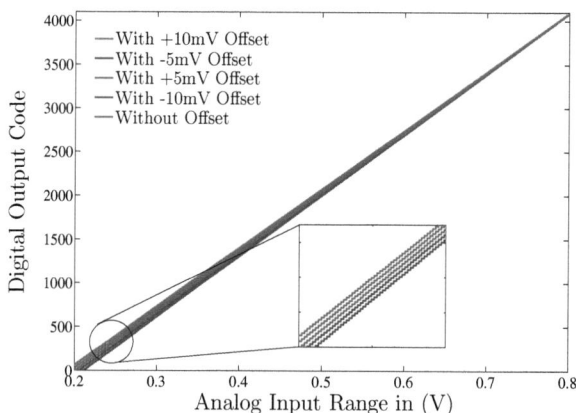

Fig. 8.4: Simulated input vs. output transfer characteristics of 12-bit SAR ADC with and without input referred offset in the buffer circuit

Effect of Aging on SAR ADC Performance

$$V_{LSB} = \frac{V_{FS}}{2^{12}}$$
$$= \frac{Vref_p - Vref_n}{2^{12}} = 146.5\mu V \qquad (8.4)$$

For example if we consider the aging induced input referred offset of 5mV in the input buffer circuit, $Vref_p = 0.8V$ and $Vref_n = 0.2V$. Then the deviation from the ideal curve at:

1. $Ain_p = 0.2V$ and $Ain_n = 0.8V$, is given by equation (8.5).

$$Dout = \frac{\frac{Vref_p - Vref_n}{2} + \frac{Ain_p - Ain_n}{2} + (5mV - 0)}{V_{LSB}} = 33, (Ideal = 0)$$
$$\Rightarrow Deviation = 33 \qquad (8.5)$$

2. $Ain_p = 0.5V$ and $Ain_n = 0.5V$, is given by equation (8.6).

$$Dout = \frac{\frac{Vref_p - Vref_n}{2} + \frac{Ain_p - Ain_n}{2} + (5mV - 2.5mV)}{V_{LSB}} = 2064, (Ideal = 2047)$$
$$\Rightarrow Deviation = 17 \qquad (8.6)$$

3. $Ain_p = 0.8V$ and $Ain_n = 0.2V$, is given by equation (8.7).

$$Dout = \frac{\frac{Vref_p - Vref_n}{2} + \frac{Ain_p - Ain_n}{2} + (5mV - 5mV)}{V_{LSB}} = 4095, (Ideal = 4095)$$
$$\Rightarrow Deviation = 0 \qquad (8.7)$$

Thus, if the worst case aging induced input referred offset of the buffer circuit is greater than $\frac{1}{2}V_{LSB} \approx 73.24\mu V$ for full-scale voltage, $V_{FS} = 0.6V_{p-p}$, then SAR ADC performance will be affected by aging degradation in the manner explained above, with magnitude depending on the offset in the input buffer.

8.3.2 Effect of Comparator Aging on SAR ADC

Aging of the comparator circuit under asymmetrical input stress for mobile phone EoL use case condition induces input referred offset of around 5mV, as presented in table 8.2. This offset value is larger than $\frac{1}{2}V_{LSB}$ of the 12-bit SAR ADC with $V_{FS} = 0.6V_{p-p}$. Therefore, the transfer characteristics depicted in figure 8.6 shows that an offset error arises in the converter due to aging induced offset in input stage of the comparator circuit. Here it is assumed that the input referred offset is not compensated using techniques like autozeroing. The deviation from ideal curve remains constant for all valid values of the input signal.

Chapter 8. Aging in Successive Approximation Register ADC

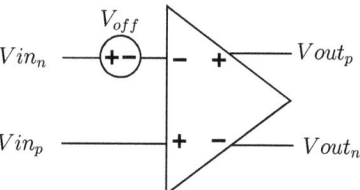

Fig. 8.5: Schematic of pre-amplifier input stage in the comparator circuit with modeled aging induced offset

For example if we consider the aging induced input referred offset of 5mV in the comparator circuit, $Vref_p = 0.8V$ and $Vref_n = 0.2V$, the deviation from the ideal curve at $Ain_p = 0.2V$ and $Ain_n = 0.8V$, is given by equation 8.8:

$$Dout = \frac{\frac{Vref_p - Vref_n}{2} + \frac{Ain_p - Ain_n}{2} + \frac{5mV}{2}}{V_{LSB}} = 17, (Ideal = 0)$$
$$\Rightarrow Deviation = 17 \tag{8.8}$$

As per the SAR conversion operation discussed in section 8.1, the offset of the comparator is sampled onto the capacitor arrays during offset cancellation and sampling phase. And is later compensated based on the principle of autozeroing discussed in section 5.2. However the accuracy of this cancellation technique depends on the open-loop gain of the input stage amplifier which is typically small for such stages. This can lead to incomplete cancellation of offset resulting into residue offset error.

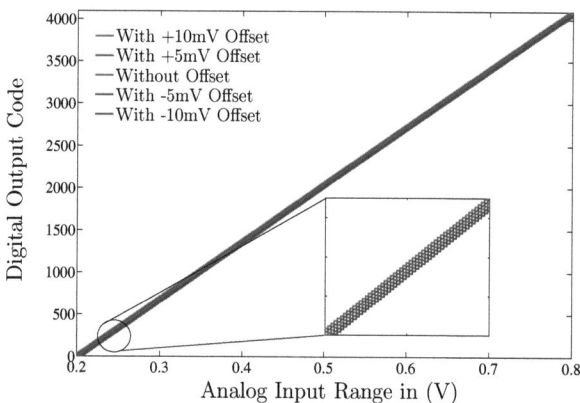

Fig. 8.6: Simulated input vs. output transfer characteristics of 12-bit SAR ADC with and without input referred offset in the comparator circuit

Effect of Aging on SAR ADC Performance

8.3.3 Combined Effect of Buffer and Comparator Aging on SAR ADC

The combined effect of aging induced offset in the input buffer and comparator circuit on the transfer characteristics of SAR ADC is depicted in figure 8.7. The deviation from ideal curve is affected based on the signs of the generated offset. For example if $Vref_p = 0.8V$ and $Vref_n = 0.2V$, the deviation from the ideal curve at $Ain_p = 0.2V$ and $Ain_n = 0.8V$ for different signs of induced offset is given by,

1. For aging induced input referred offset of +5mV in both input buffer and comparator circuits,

$$Dout = \frac{\frac{Vref_p - Vref_n}{2} + \frac{Ain_p - Ain_n}{2} + 5mV + \frac{5mV}{2}}{V_{LSB}} = 50, (Ideal = 0)$$

$$\Rightarrow Deviation = 50 \qquad (8.9)$$

2. For aging induced input referred offset of +5mV in input buffer circuit and −5mV in comparator circuits,

$$Dout = \frac{\frac{Vref_p - Vref_n}{2} + \frac{Ain_p - Ain_n}{2} + 5mV - \frac{5mV}{2}}{V_{LSB}} = 16, (Ideal = 0)$$

$$\Rightarrow Deviation = 16 \qquad (8.10)$$

Therefore offsets induced in opposite directions compensate the aging mechanisms induced degradation to great extent.

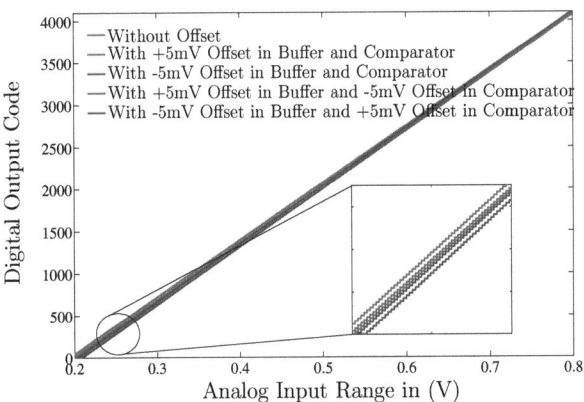

Fig. 8.7: Simulated input vs. output transfer characteristics of 12-bit SAR ADC with and without input referred offset in the input buffer and comparator circuit combined

90   Chapter 8. Aging in Successive Approximation Register ADC

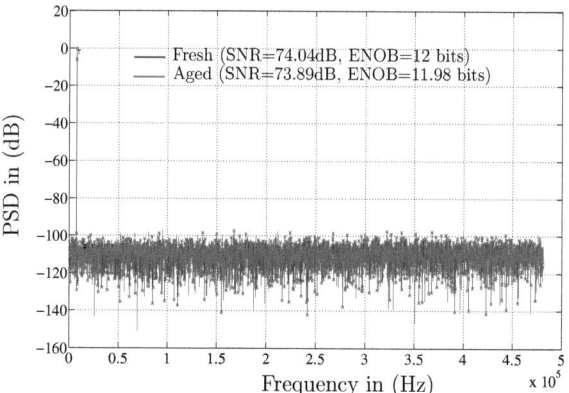

Fig. 8.8: **Power Spectral Density plot of aged 12-bit SAR ADC circuit with 10mV input referred offset in the input buffer circuit compared with fresh ADC**

The simulated differential non-linearity (DNL) and integral non-linearity (INL) for an aged 12-bit SAR ADC under mobile phone EoL use case stress condition are shown in figure 8.9. These performances are not affected due to aging of the input buffer and comparator circuits. This results from the fact that gain and offset errors are corrected before evaluating INL and DNL of a Nyquist rate ADC. It was confirmed by simulations that also the spectral characteristics of the ADC were not affected. The resultant ENOB is 11.98 bits, which represents that the linearity of this SAR ADC is not affected by the aged buffer circuit as illustrated in figure 8.8. There are proven methods to correct gain

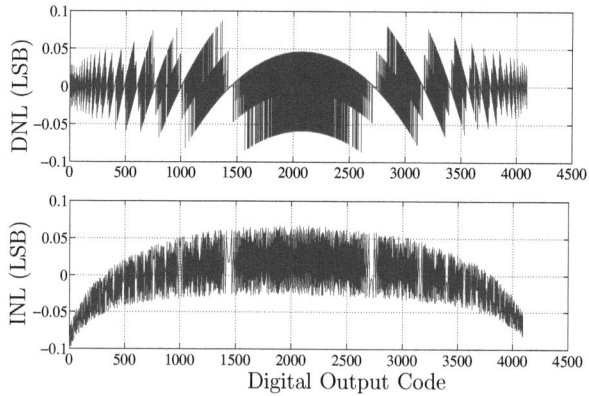

Fig. 8.9: **Simulated DNL and INL of aged 12-bit SAR ADC with 10mV input referred offset in the input buffer circuit**

and offset errors. But it is important to note that these errors will vary over time. Hence special countermeasures need to be implemented to guarantee a stable and correct circuit function for the whole lifetime of the circuit.

8.4 Countermeasures

Active countermeasures like chopper stabilization and autozeroing techniques were discussed in chapter 5 as very effective techniques in mitigating effects of aging degradation in OTA circuits. Apart from these techniques, digital background calibration techniques are also used quite frequently to overcome effects of non-idealities in high resolution state-of-the-art ADC circuits [78]. Digital calibration is preferred over analog calibration due to high robustness and integration density together with low cost and power requirements of digital circuitry. They are implemented to calibrate errors resulting from variation in mismatch over temperature, power supply or aging degradation [79]. They perform without interruption of the normal ADC operation. In contrast foreground calibration is done when the ADC is not converting (e.g. immediately after power ON) and thus cannot react to continuous error changes unless the ADC is interrupted and calibrated again.

The digital background calibration process can be divided into two phases; first during the measurement phase the deviation of the ADC output with respect to an ideal converter is estimated in the digital domain. And in the second phase, the evaluation phase, in order to calibrate, the estimation of the errors from the measurement phase is subtracted from the raw output code of ADC [80,81].

Digital background calibration techniques rely on adaptive algorithms in order to minimize the error in the estimation of the non-idealities in the ADC [4]. Adaptive algorithms are used in two different approaches: channel error identification [82] and correlation-based [83] techniques. The first approach requires an accurate reference ADC, e.g. a sigma-delta converter, as reference ADC. Correlation based calibration is a statistical approach to estimate and calibrate errors digitally and has the advantage that minimal extra analog design effort is required for calibration purposes, i.e. there is no redundant hardware so that the extra implementation is kept minimal. A disadvantage for this implementation can be the long correlation time, i.e. the calibration of the ADC can last for around one minute [84].

The effectiveness of correlation-based digital background calibration techniques in overcoming aging induced performance degradation in SAR ADC circuits needs to be investigated.

8.5 Summary

In this chapter aging degradation in high performance Nyquist rate SAR ADC was discussed and analyzed. The impact of aging on building blocks of 12-bit SAR ADC viz., input buffer and comparator, and its individual and combined effect on ADC performance was evaluated under asymmetrical input stress condition. Using the findings related to aging degradation in closed loop OTA (buffer) and open loop OTA (comparator input stage) configurations, the impact of its aging on the performance of a SAR ADC circuit was evaluated. The most severely affected performance due to aging under asymmetrical stress in buffer and comparator circuit configurations was found to be offset.

It was shown that the aging induced offset in the input buffer stage introduces gain error, whereas aging induced offset in the comparator stage introduces offset error in the ADC transfer characteristic. The combined effect results into time varying gain and offset errors in its transfer characteristic. If this offset in the buffer and comparator stage have opposite signs then they tend to compensate the aging mechanisms induced performance degradation to a considerable extent.

The analysis carried out in this chapter leads to the conclusion that for mobile phone EoL use case condition the performance of 12-bit SAR ADC implemented in 32nm high-κ metal gate CMOS technology, is not severely affected by aging degradation. However, for other stress conditions leading to higher values of degradation there is a need to implement special countermeasures which can correct time varying errors resulting from stress induced aging degradation in high resolution ADC's implemented in nano-scale CMOS technology.

Chapter 9
Aging in Sigma Delta ADC

Sigma delta ($\Sigma\Delta$) ADC implements a data conversion technique, in which high performance is achieved by both oversampling and noise shaping. In discrete time $\Sigma\Delta$ ADC different functionalities are implemented using switched capacitor circuits. One of the main advantages of using $\Sigma\Delta$ ADC is that high accuracy can be realized at relaxed analog circuit requirement. In this chapter the effect of aging degradation mechanisms on the performances of state-of-the-art fully differential third-order, 2-stage, multi-bit (17-level) $\Sigma\Delta$ ADC are evaluated and discussed.

9.1 Introduction to Sigma Delta ADC

The application field of $\Sigma\Delta$ ADC's is between low to medium speed. The high oversampling ratio (OSR) (larger than 64) combined with single-bit quantizer provides very good resolution (over 80dB signal to noise and distortion ratio (SNDR)), but it is limited only to low signal bandwidth. In order to expand the signal bandwidth, the OSR must be reduced, e.g. down to 8 to 16. But at low OSR high resolution can hardly be guaranteed due to weak noise shaping effect. If accuracy is still desired, the in-band quantization noise must be further suppressed. For this reason, multi-bit quantizer is introduced. In fact, after employing multi-bit quantizer, the quantization noise reduction is caused by smaller quantization step size. Additionally, the stability is also better than single-bit quantizer, since implementation of larger scaling coefficients is possible.

Apart from increasing the resolution of quantizer, using more aggressive noise-shaping function is another approach to enhance SNDR. Actually this can be done by employing $\Sigma\Delta$ ADC with higher order quantization noise transfer function (NTF). Theoretically, the higher the order of NTF, the better the noise shaping effect will be. But increasing the orders of $\Sigma\Delta$ ADC results in stability problems and system overload may occur even when the input signal is not quite large. Fortunately the root-locus simulation shows, if the coefficients of the integrator are properly selected, first and second order sigma-delta converters are intrinsically stable [85]. By using cascaded topologies or multi-stage-noise-shaping (MASH) topologies, in which high-order is accomplished by interconnecting two

94   Chapter 9. Aging in Sigma Delta ADC

or more first and/or second order sigma-delta converters, the stability problem in high-order $\Sigma\Delta$ ADC is avoided [86].

9.1.1 Sigma Delta ADC Implementation

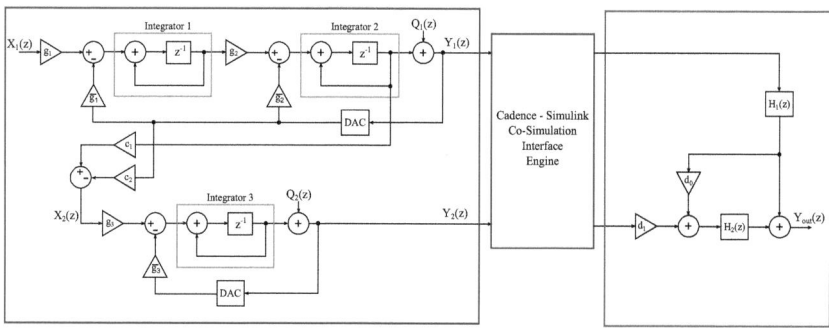

Fig. 9.1: 3rd order multi-bit (17-level) 2-1 MASH $\Sigma\Delta$ ADC model

A third-order 2-stage cascaded $\Sigma\Delta$ ADC with multi-bit (17-level) quantizer, satisfying both the high-speed and high-resolution requirements is investigated for performance degradation under aging effect. The circuit comprises two major portions, the analog noise-shaping circuit and digital quantization error cancellation logic as illustrated in figure 9.1. A fully differential implementation of this circuit is done in Cadence and Simulink environment. The analog noise-shaping circuit in the left (blue block) is implemented in Cadence environment using 32nm high-κ metal gate CMOS technology except for the multi-bit quantizer and digital to analog converter (DAC). These two blocks are modeled using Verilog-A. And the digital logic to cancel the quantization noise of first stage is implemented in Simulink. The third-order noise-shaping is realized in two stages by interconnecting second-order and first-order $\Sigma\Delta$ converters, where the quantization error of first stage is shaped by the second stage. One of the most essential portions of cascaded architecture is the digital quantization error cancellation logic, where theoretically the quantization error of first stage can be completely canceled out and the quantization error of second stage is first shaped by the second stage, and then further shaped by the second-order high-pass filter $H_2(z)$ in quantization error cancellation logic. As a result, the output signal (Y_{out}) should contain only two components, the delayed converted input signal and shaped quantization error of second stage. Through employing multi-bit (17-level) quantizer and DAC in both the stages, an aggressively increased interstage gain, namely c_1 and c_2, can be implemented without overloading the modulators.

Introduction to Sigma Delta ADC   95

The output of the first and second stage is given by:

$$Y_1(z) = z^{-2} \cdot X_1(z) + (1 - z^{-1})^2 \cdot Q_1(z)$$
$$Y_2(z) = z^{-1} \cdot X_2(z) + (1 - z^{-1}) \cdot Q_2(z) \qquad (9.1)$$
$$where, X_2(z) = c_1[Y_1(z) - Q_1(z)] - c_2 \cdot Y_1(z)$$

Therefore, the digital output $Y_1(z)$ and $Y_2(z)$ in terms of $X_1(z)$, $Q_1(z)$ and $Q_2(z)$ are given by:

$$Y_1(z) = z^{-2} \cdot X_1(z) + (1 - z^{-1})^2 \cdot Q_1(z)$$
$$Y_2(z) = z^{-1} \cdot [(c_1 - c_2) \cdot Y_1(z) - c_1 \cdot Q_1(z)] + (1 - z^{-1}) \cdot Q_2(z) \qquad (9.2)$$

If $c_1 = c_2$ then,

$$Y_1(z) = z^{-2} \cdot X_1(z) + (1 - z^{-1})^2 \cdot Q_1(z)$$
$$Y_2(z) = -c_1 \cdot z^{-1} \cdot Q_1(z) + (1 - z^{-1}) \cdot Q_2(z) \qquad (9.3)$$

Now, the cancellation of the quantization error of the first stage is performed using:

$$Y_{out}(z) = Y_1(z) \cdot z^{-1} + Y_2(z) \cdot \frac{1}{c_1} \cdot (1 - z^{-1})^2 \qquad (9.4)$$

$$Y_{out}(z) = z^{-3} \cdot X_1(z) + \frac{1}{c_1} \cdot (1 - z^{-1})^3 \cdot Q_2(z) \qquad (9.5)$$

The optimal coefficients for this third-order 2-stage cascaded multi-bit $\Sigma\Delta$ ADC are listed as following:

$$g_1 = 0.5, \overline{g_1} = 0.5, g_2 = 2, \overline{g_2} = 2, g_3 = 1, \overline{g_3} = 1,$$
$$c_1 = 8, c_2 = 8, d_0 = 0, d_1 = 0.125 \qquad (9.6)$$

The maximum achievable SNDR is given by equation (9.7) [87]:

$$SNDR_{max} = 1.76 + 6.02 \cdot m + (20 \cdot L + 10).log_{10}(OSR) - 10 \cdot log_{10} \cdot (\frac{\pi^{2 \cdot L}}{2 \cdot L + 1}) \qquad (9.7)$$

where m = Number of quantizer bits and L = order of $\Sigma\Delta$ ADC

According to equation (9.7), the maximum SNDR of a third-order 17-level $\Sigma\Delta$ ADC at the full scale input amplitude and $OSR = 16$ is given by:

$$SNDR_{max} = 1.76 + 6.02 \cdot 4.123 + (20 \cdot 3 + 10) \cdot log_{10}(16) - 10 \cdot log_{10} \cdot (\frac{\pi^{2 \cdot 3}}{2 \cdot 3 + 1})$$
$$= 89.12 dB \qquad (9.8)$$

The circuit schematic of analog noise-shaping block in the fully differential third-order, 2-stage, 17-level, cascaded sigma-delta ADC converter implemented in Cadence

96 Chapter 9. Aging in Sigma Delta ADC

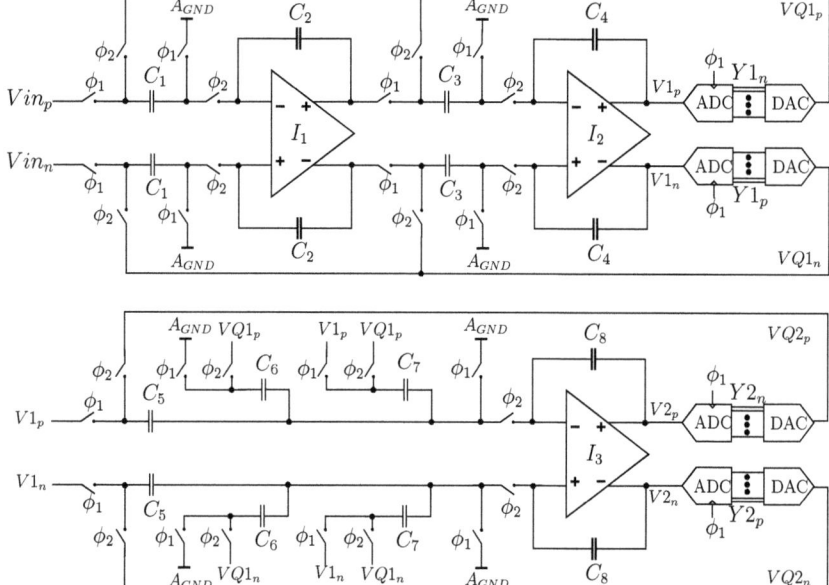

Fig. 9.2: Schematic of analog noise-shaping circuit in 3rd order multi-bit (17-level) 2-1 MASH ΣΔ ADC

environment is illustrated in figure 9.2. Here the multi-bit quantizer (ADC) and DAC models are implemented with Verilog-A. The digital quantization noise cancellation circuit is implemented in Simulink environment. The Cadence-Simulink co-simulation is performed with an interface engine which enables bi-directional flow of data between the two environments during simulation. The complete system was simulated with:

Input signal bandwidth: $25KHz$
Input signal frequency: $10.15625KHz$
Input signal amplitude (V_{p-p}): $0.6V$
Oversampling ratio (OSR): 16
Sampling frequency: $800KHz$
Number of FFT points used to calculate SNDR: $2^{10} = 1024$

The simulated SNDR for the fresh circuit is evaluated to be $91.62dB$. It is higher than that evaluated in equation (9.8) since equation (9.7) does not account for the additional noise shaping achieved by the factor $\frac{1}{c_1}$ in equation (9.5). The effective number of bits (ENOB) was calculated to be 14.93 bits.

Aging in Sigma Delta ADC Building Blocks   97

9.2 Aging in Sigma Delta ADC Building Blocks

This section presents an evaluation of aging degradation in the $\Sigma\Delta$ ADC building blocks, particularly the integrator, multi-bit quantizer and DAC circuits. The impact of individual and combined degradation of different building blocks on overall ADC performance is discussed separately. The aging induced performance degradation in the closed and open loop operational transconductance amplifier (OTA) configurations for a mobile phone EoL use case stress condition discussed in chapter 4 are added to the $\Sigma\Delta$ ADC building blocks. The impact of switch aging is neglected since proper device sizing can take care of the aging induced increase in switch "ON" resistance as discussed in chapter 7.

9.2.1 Effect of Integrator Aging on Sigma Delta ADC

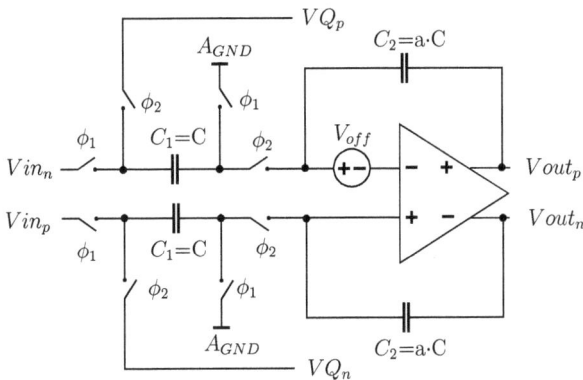

Fig. 9.3: Switched capacitor integrator stage with scaling and subtraction circuit

The third-order, 2-stage $\Sigma\Delta$ ADC requires two modulators implemented using three switched capacitor integrators. The integrator circuit basically consists of OTA along with few switches and capacitors. One such fully differential switched capacitor integrator stage with scaling and subtraction circuit is illustrated in figure 9.3. During the sampling phase ϕ_1 is "ON" and ϕ_2 is "OFF", allowing the voltages across capacitors C_1 to track Vin while the charge from the previous cycle is stored on capacitors C_2 connected across an OTA circuit. Next during the integration phase ϕ_1 is "OFF" and ϕ_2 is "ON", voltage stored across capacitors C_1 is subtracted by voltage VQ, where VQ is the 17-level quantized output signal converted back to analog domain using 17-level DAC. The remaining charge is transferred to capacitors C_2 through virtual ground node of the OTA circuit. Since $C_2 = a \cdot C_1$ the voltage stored on C_2 is scaled by factor $\frac{1}{a}$, i.e. $Q_{C_2} = \frac{1}{a} \cdot C \cdot (Vin - VQ)$.

Since the OTA in the integrator circuit always operates in the closed loop configuration, based on the findings in section 4.1.1 an input referred offset of 0.05mV is added to

Offset (mV)			SNDR (dB)	ENOB (bits)
Integrator1	Integrator2	Integrator3		
0	0	0	91.62	14.93
0.05	0	0	76.3	12.38
0	0.05	0	88.3	14.38
0	0	0.05	90.9	14.81
0.05	0.05	0.05	76.0	12.33

Table 9.1: **Impact of integrator aging degradation on ΣΔ ADC performance**

each of the OTAs in the three integrator stages. The sensitivity of ΣΔ ADC performance (SNDR) towards this aging induced offset in each of the integrator stage is evaluated independently by applying offset to only one the three OTA circuits at a time. Finally a combined effect of aging degradation of all integrator stages on the ΣΔ ADC performance is evaluated. The power spectral density (PSD) plots for the implemented third order ΣΔ ADC with offset modeled in different integrator stages are illustrated in figure 9.4. The simulation results evaluating impact of integrator aging degradation on ΣΔ ADC performance are summarized in table 9.1.

The results illustrated in figure 9.4 and summarized in table 9.1 confirm that the ΣΔ ADC performance is most sensitive to the aging induced input referred offset in the first integrator stage compared to the similar offset in the second and third integrator stages. Here an input referred offset of 0.05mV in the first integrator stage (integrator1) degrades the SNDR by around 15dB whereas the same offset in second and third integrator stage (integrator2 and integrator3) only slightly affects the SNDR (decrease by 3dB and 1dB

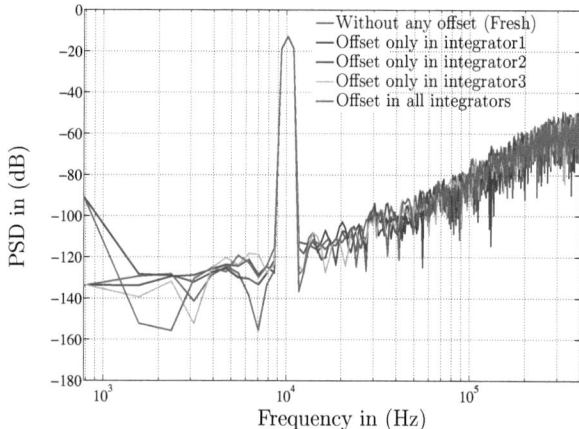

Fig. 9.4: **Comparison between PSD's of third order ΣΔ ADC's with aging induced input referred offset in different integrator stages**

respectively). Thus the offset in the first integrator stage outweighs by far the impact of aging on sigma-delta performance degradation, leading to a large SNDR drop down to 76.3dB. The increased degradation of SNDR comes from the increase in DC power in the $\Sigma\Delta$ ADC's PSD which is visible in figure 9.4.

9.2.2 Effect of Multi-Bit Quantizer Aging on Sigma Delta ADC

To realize the required 17-level multi-bit quantizer in both stages of the fully differential third-order $\Sigma\Delta$ ADC, two differential pairs i.e. total four flash ADC's are implemented. A differential pair of this 17-level flash ADC circuit used for each stage of the $\Sigma\Delta$ ADC is illustrated in figure 9.5. The 16-bit output of these flash ADC's are thermometer coded and are fed to a 16-bit thermometer coded DAC which will be discussed in next section. The primary source of error due to degradation in the flash ADC comes from aging induced offset in the input pre-amplifier stage of the comparator circuits. As discussed in chapter 4.1.2 significant input referred offset is generated due to aging in the open loop OTA circuit configuration under asymmetrical stress conditions. Each flash ADC consist of 16 comparator circuits and 16 reference voltages as illustrated in figure 9.6. The negative input of the $comparator_i$ is connected to the resistor ladder providing the reference voltage $Vref_i$, where $i\,[\in 0..15]$ and these voltages are fixed. Therefore, higher the value of $Vref_i$ of the corresponding $comparator_i$, the more it's input transistor device connected to the negative input will be stressed due to higher V_{gs}. The positive input of all the comparators are connected to input voltage Vin. Therefore all input transistors connected to the positive terminal will experience the same level of stress. Hence for $Vin = 0V$ worst case positive offset is induced in $comparator_{15}$ followed by $comparator_{14}$ and so on. On the other hand for $Vin = V_{FS}$ worst case negative offset is induced in

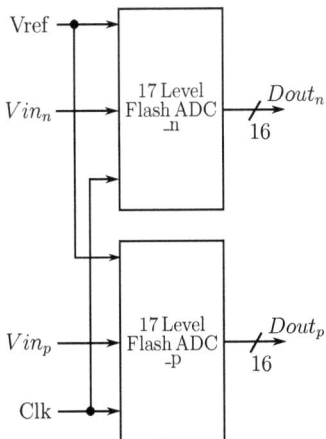

Fig. 9.5: Differential configuration of 17-level flash ADC used as quantizer in multi-bit $\Sigma\Delta$ ADC

100   Chapter 9. Aging in Sigma Delta ADC

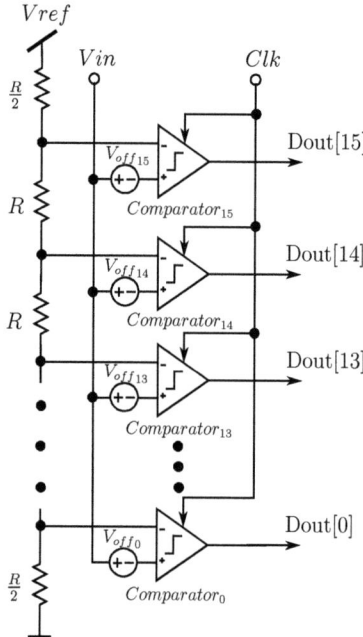

Fig. 9.6: 17-level flash ADC circuit

$comparator_0$ followed by $comparator_1$ and so on.

This $\Sigma\Delta$ ADC is implemented in the fully differential configuration, hence for asymmetrical stress condition all the comparators in one of the flash ADC from the differential pair experience positive offset where as the others experience negative offset. For e.g. considering the mobile phone EoL use case condition under asymmetrical input stress ($Vin_n = 0V$ and $Vin_p = 1.155V$) for the implemented $\Sigma\Delta$ ADC, the comparators in the flash ADC (_n) in the negative path with input Vin_n generate maximum positive input referred offset at $comparator_{15}$ followed by $comparator_{14}$ and so on. And at the same time the comparators in the flash ADC (_p) in the positive path with input Vin_p generate maximum negative input referred offset at $comparator_0$ followed by $comparator_1$ and so on. These different values of aging induced offset are modeled in all the flash ADC's implemented using Verilog-A for both stages of the $\Sigma\Delta$ ADC and the simulation results are summarized in table 9.2. The offset values used in this table are not determined by aging simulation, but are used only for demonstration purpose.

The resolution of the 17-level (≈ 4.123 bits) flash ADC is given by $\frac{1}{2}V_{LSB} \approx 28.5mV$ for $V_{FS} = 1V_{p-p}$. Thus for input referred offsets less than $28.5mV$ there are no missing codes and hence no significant degradation of $\Sigma\Delta$ ADC performance is expected. This is confirmed by simulation results presented under table 9.2 where for data Set 2 small

Aging in Sigma Delta ADC Building Blocks 101

Comparator No.	Set 1		Set 2		Set 3	
	_n(mV)	_p(mV)	_n(mV)	_p(mV)	_n(mV)	_p(mV)
15	0	0	14	0	30	0
14	0	0	12	0	25	0
13	0	0	10	0	20	0
12	0	0	8	0	15	0
11	0	0	6	0	10	0
10	0	0	4	0	5	0
9	0	0	2	0	1	0
8	0	0	0	0	0	0
7	0	0	0	0	0	0
6	0	0	0	−2	0	−1
5	0	0	0	−4	0	−5
4	0	0	0	−6	0	−10
3	0	0	0	−8	0	−15
2	0	0	0	−10	0	−20
1	0	0	0	−12	0	−25
0	0	0	0	−14	0	−30
SNDR (dB)	91.62		88.1		84.7	
ENOB (bits)	14.93		14.35		13.78	

Table 9.2: **Impact of 17-level flash ADC aging degradation on $\Sigma\Delta$ ADC performance**

SNDR degradation is noted and for Set 3 the SNDR is reduced down to 84.7dB.

9.2.3 Effect of Current Steering DAC Aging on Sigma Delta ADC

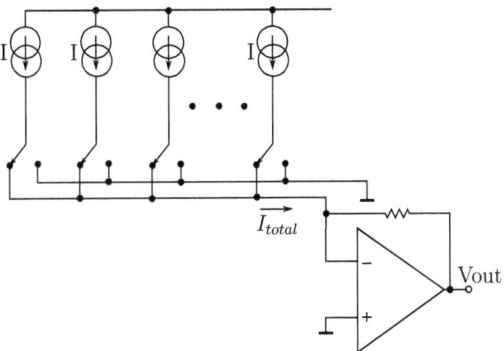

Fig. 9.7: 17-level current steering DAC circuit

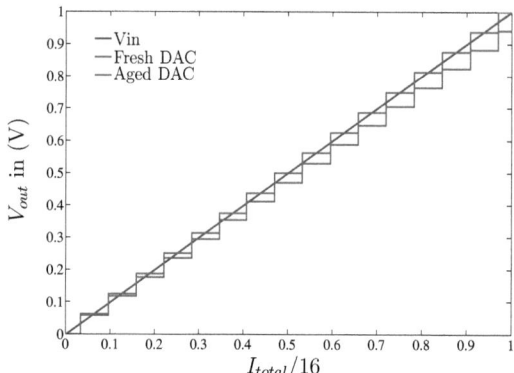

Fig. 9.8: Comparison between fresh and aged DAC transfer characteristics

In case of high data rate requirements commonly current steering DAC circuits are implemented in multi-bit $\Sigma\Delta$ ADC. The unit elements of the thermometer coded DAC circuit are implemented using 16 identical unit current sources (I) as illustrated in figure 9.7, which are realized by current mirroring a reference current (I_{ref}) into these current sources. The contribution of these current sources to the total output current (I_{total}) and therefore output analog voltage (V_{out}) is controlled by the switches driven by the digital inputs.

Due to the fact that all the unit current sources witness identical bias conditions hence they are stressed equally. Thus the shifts in transistor parameters due to aging degradation are equal for all unit current sources. For this reason, degradation mechanisms are more likely to generate gain errors (G_{DAC}) in the DAC transfer characteristics than nonlinear errors as illustrated in figure 9.8 [9]. Moreover, the aging induced input referred offset in the buffer circuit will result into offset error in DAC transfer characteristics. Different gain errors ($1V_{LSB}$ and $2V_{LSB}$) in the DAC transfer characteristics and offset error of (0.05mV) in the buffer circuit are modeled in the DAC implemented using Verilog-A and the simulation results of $\Sigma\Delta$ ADC performance degradation are summarized in table 9.3. From simulation results it can be seen that even a gain error of $2V_{LSB}$ does not affect the SNDR significantly when no asymmetry is introduced in the fully differential circuit due to aging degradation. ADC performance could be affected in the

Gain Error (V_{LSB})	Offset (mV)	SNDR (dB)	ENOB (bits)
0	0	91.62	14.93
1	0	89.6	14.59
2	0	88.5	14.40
2	0.05	85.5	13.90

Table 9.3: Impact of 17-level DAC aging degradation on $\Sigma\Delta$ ADC performance

case of asymmetry in the aging induced gain errors since the DAC outputs are fed back to be subtracted from the original input signal and therefore undergo the same transfer function as the input signal itself [88].

9.2.4 Combined Effect of integrator, Quantizer and DAC Aging on Sigma Delta ADC

Gain Error (LSB)	Offset (mV)	SNDR (dB)	ENOB (bits)
0	0	91.62	14.93
2	0.05	75.9	12.32

Table 9.4: Combined effect of building blocks aging degradation on $\Sigma\Delta$ ADC performance

The combined effect of aging induced degradation in the integrator, 17-level quantizer and 17-level DAC stages on the performance of $\Sigma\Delta$ ADC is evaluated in this section. An aging induced input referred offset of 0.05mV in all the OTA's of the integrator stages is modeled which degrades the SNDR to 76dB as discussed in section 9.2.1. Positive and negative aging induced offset in the comparator stages of differential quantizers implemented using flash ADC is modeled using the offset values from the Set 3 of table 9.2. And the aging induced gain error of 2LSB and offset error of 0.05mV due aging of the buffer circuit, in transfer characteristics of the current steering DAC are also modeled. The combined effect of aging of these building blocks of $\Sigma\Delta$ ADC in summarized in table 9.4. The SNDR is reduced to 75.9dB which highlights that the increase in DC power in the PSD of the $\Sigma\Delta$ ADC resulting from the aging induced offset in the first integrator stage dominates the overall degradation.

9.3 Countermeasures

The complete cancellation of the quantization error of the first $\Sigma\Delta$ stage done by the digital quantization error cancellation logic shown in figure 9.1 relies on the perfect matching of the analog NTF of the first stage (NTF_1) and its digital counterpart $H_2(z)$ [89]. While the digital circuitry will intrinsically provide the desired transfer functions, the analog transfer functions are affected not only by non-idealities but also by aging induced parameter drifts which will lead to the so called noise-leakage. This means that the quantization noise of the first stage cannot be totally removed by the cancellation logic and therefore leaks through to the $\Sigma\Delta$ ADC output (Y_{out}) [90]. This implies, that the noise-leakage might be reduced by adapting the digital transfer function $H_2(z)$ to the modified analog transfer function NTF_1 by estimating new NTF_1 itself. Therefore the aging degradation induced impacts on the analog transfer functions could be corrected [88]. The increase of DC power in the $\Sigma\Delta$ ADC's PSD needs to be handled by post processing its output. The effectiveness of correlation-based digital background calibration techniques discussed in section 8.4, in overcoming other aging induced performance degradation in $\Sigma\Delta$ ADC circuits needs to be investigated.

9.4 Summary

Oversampling sigma delta ADC is relatively robust in terms of accuracy of its matched analog components compared to Nyquist rate ADC. The investigations carried out in this chapter related to the impact of aging degradation on the performance of $\Sigma\Delta$ ADC revealed the fact that, to a high extent the precision of the $\Sigma\Delta$ circuits exceeds the precision of its components. The increase of the DC power in the PSD related to the offset induced due to aging in the first integrator stage mainly affected the ADC performance. The impact of aging degradation in the multi-bit quantizer and the DAC circuit was not significant due to its low resolution.

Any mismatch induced in aging behavior was not accounted due to the use of typical analog size transistors used in the implementation of these building block of the $\Sigma\Delta$ ADC. Investigations related to the errors induced due to this mismatch and also dynamic errors due to relaxation behavior of the BTI degradation could give additional insight into the impact of aging on the performance of $\Sigma\Delta$ ADC.

Chapter 10
Conclusions and Outlook

10.1 Conclusion

In this research analog and mixed signal circuits designed in advanced state-of-the-art 32nm high-κ metal gate CMOS technology were investigated for performance degradation resulting from aging wearout mechanisms. An aggressive non-constant field scaling in the deep-submicrometer CMOS technology, introduction of new dielectric material like high-κ and increase in operating temperatures due to high density of transistors per chip has inevitably led to rising reliability concerns from degradation mechanisms such as bias temperature instability (BTI) and hot carrier injection (HCI). A combined effect of these degradation mechanisms on the performance of analog and mixed signal circuits was evaluated analytically, by simulations and by measurements on test hardware. The worst case stress conditions for different circuit topologies were studied and weakest spots in the circuit susceptible to highest aging degradation were located. Instead of using unrealistic elevated stress conditions application specific qualification was carried out considering the application conditions including temperatures and voltage ranges the electronic system is going to withstand during its lifetime. In this investigation all the evaluations were carried out for an end-of-life (EoL) mobile phone use case stress condition. Accelerated test conditions mapping accurately to this end-of-life use case conditions were used ensuring that no new aging mechanism was introduced while using accelerated stress. This was confirmed by comparing simulation and measurement results. Efficient countermeasures to compensate aging induced performance degradation were introduced and demonstrated using measurement results. Significant reduction in performance degradation was achieved using these on-line monitoring and compensation techniques.

Asymmetrical stress induces mismatch in matched transistor pairs. Hence, while evaluating performance degradation in differential analog and mixed signal circuits, aging induced mismatch in matched pairs is more important compared to other individual transistor's parameter drift. Therefore, asymmetrical input stress is most harmful for reliability of such circuits. In fully differential operational amplifier (OTA) circuits aging under asymmetrical stress induces offset, whereas other performances like amplifier gain, bandwidth and phase margin were not considerably affected provided all the transistors remain in saturation after aging. The performance of the OTA in open loop configuration is significantly more degraded compared to that in closed loop configuration. The methodology for analytical evaluation of aging degradation in linear circuits revealed that the transistor most affected by aging degradation is not always the weakest spot concerning reliability in circuits. The differential circuit is most affected by the mismatch in the matched transistor pair toward which the performance under investigation has highest sensitivity. Further, the comparison between two operational amplifier topologies showed that simple Miller OTA was more affected by aging as compared to the folded cascode OTA topology. Performance of folded cascode OTA topology was more robust to aging degradation because of shielding of its transistors from high bias voltages by the cascode structures.

Active countermeasures are required to mitigate aging induced performance degradation in highly precise and accurate AMS circuits. Two on-line techniques viz., chopper stabilization (CHS) and auto zeroing (AZ) were proposed and evaluated. Using CHS technique, a significant reduction of more than 96% was measured in the aging induced offset of the OTA test chips. Further, relaxation of the offset due to BTI was not observed due to symmetrical degradation and differential signaling. AZ was also very effective in offset cancellation, however the accuracy of the technique is limited by the open loop gain of the OTA circuit and its application is restricted to sampled data systems.

The performance of ring oscillator circuit implemented with minimum gate length devices in 32nm technology node is affected by relatively high conducting hot carrier injection (CHCI) mechanism compared to BTI mechanism under AC stress for mobile phone EoL use case condition. Hence lifetime enhancement under AC stress is no longer given. An effective monitoring and background compensation technique to counteract aging induced performance degradation in ring oscillator circuit was demonstrated by measurement results. The adaptive and bipolar selection of compensation voltage steps enhances the tracking algorithm which was able to reduce the performance degradation by 98%. This tracking algorithm could further be implemented on-chip to monitor and compensate aging degradation on-the-fly. Also fast recovery behavior could be monitored with this concept.

CMOS switches most commonly used in switched capacitor circuits experience increase in their "ON" state drain to source resistance (R_{on}) due to aging degradation. This results into incomplete transfer of charge onto capacitors and hence circuit performance degradation. The degradation effect is worst at minimum overdrive voltages and maximum operating frequency. A switch aged for an EoL mobile phone use case condition under DC stress experiences 19% increase in its R_{on} when charaterized at minimum overdrive voltage post aging. Under AC stress, around 50% reduction in aging degradation due to recovery of BTI mechanism was measured compared to that under DC stress. The ineffectiveness of using bipolar stress as on-line countermeasure to compensate aging degradation using accumulation stress was discussed. And hence the importance of proper sizing of switches to ensure reliable working of switched capacitor circuits at worst case overdrive, maximum frequency and under aging degradation was highlighted.

Aging induced degradation of both high performance Nyquist rate and oversampling ADC circuits were analyzed and evaluated. In a 12-bit successive approximation (SAR) ADC aging induced offset in the input buffer stage introduces gain error, whereas aging induced offset in the comparator stage introduces offset error in the ADC transfer characteristic. The combined effect results into time varying gain and offset errors in its transfer characteristic. If this offset in the buffer and comparator stage have opposite signs then they tend to compensate the aging mechanisms induced performance degradation to a great extent. The analysis carried out on a 12-bit SAR ADC implemented in 32nm high-κ metal gate CMOS technology leads to the conclusion that for mobile phone EoL use case condition, its performance is not severely affected by aging degradation if the aging induced offsets, particularly in the comparator circuit, are handled using countermeasure techniques. Similarly, for the oversampling sigma-delta ($\Sigma\Delta$) ADC which is considered to be more robust against non-idealities in its analog building blocks, investigations related to aging degradation revealed that the impact of aging on its performance under mobile phone EoL use case condition is also not very severe. A third-order 2-stage cascaded $\Sigma\Delta$ ADC with multi-bit (17-level) quantizer was investigated for performance (SNDR) degradation under aging effect. The aging induced offset in the first modulator degrades the SNDR by 16% mainly due to the increase of the DC power in the PSD. The impact of aging degradation in the multi-bit quantizer and the DAC circuit was not significant due to its low resolution.

Based on these investigations it can be concluded that the wearout mechanisms in 32nm high-κ metal gate CMOS technology are not show-stoppers for the development of analog and mixed signal systems. However a careful analysis of aging effects at device and circuit level, right from the design phase and incorporation of effective countermeasures is necessary while implementing highly accurate and precise circuits.

10.2 Outlook

10.2.1 Variability in Aging Degradation

BTI and HCI induced degradation in analog (micrometer) sized transistors are typically deterministic in nature and are also called temporal deterministic unreliability effects. In this case identical parameter shift is induced in matched transistor pairs on application of symmetrical stress. This was observed from the very small aging induced input referred offset ($11\mu V$) measured in the Miller OTA circuit with chopper stabilization technique. Here all the matched transistor pairs in the OTA circuit were stressed with symmetrical stress. However scaling down the transistors to nanometer dimensions changes the deterministic nature of degradation effect to stochastically distributed effect, e.g. due to varying number of traps in the oxide and the interface to the channel. This is now termed as temporal stochastic unreliability effects. This results in time dependent shift in parameters of the transistors combined with time dependent increase in standard deviation on these parameters [91]. Hence the matched pairs can develop mismatch during operating lifetime even under symmetrical stress and lead to circuit failures. In this work mismatch induced due to process variation (e.g. threshold voltage mismatch due to statistical variation of the number of doping atoms in a transistor) and variation in aging degradation were not considered and hence require further investigations.

10.2.2 BTI Recovery Effect

There have been tremendous efforts in understanding the mechanisms and physics behind BTI recovery behavior since it has been the source of disagreements and confusions related to BTI measurement issues [45, 49, 92]. However, at the moment no generally accepted models are available for evaluating the recovery behavior of BTI degradation under AC stress with arbitrary stress/recovery sequences at EoL and during operating lifetime of the MOSFET device and circuit. Hence there are very limited investigations carried out related to the impact of transient threshold voltage (V_{th}) change resulting due to the BTI recovery effect on analog and mixed signal circuits performance [93]. This transient V_{th} changes can induce dynamic errors in precise and accurate analog and mixed signal circuits and hence needs to be reviewed. Also, variability in the recovery behavior of BTI mechanism is not investigated fully yet.

10.2.3 Novel Devices and Design Strategies

As the conventional CMOS planar structure is approaching its physical limits, the International Technology Roadmap for Semiconductors (ITRS) [94] reflects the trend of migration from geometrical scaling to equivalent scaling and design equivalent scaling in the semiconductor industry. Equivalent scaling refers to the performance improvement with introduction of new material, new device structures (e.g. 3-D devices) and other non-geometrical process innovations. Design equivalent scaling represents performance improvement by innovative design, improved software and data processing. Recent experimental investigations show that the new devices like multiple gate field-effect transistor (MuGFET) with standard orientation exhibits worse BTI degradation effects than planar devices due to higher availability of $Si - H$ bonds at the fin sidewalls and due to the self-heating effect caused by the silicon-on-insulator (SOI) body [95]. Fin-shaped field effect transistor (FinFET) devices are found to have improved HCI immunity however it significantly depends on factors like interface state generation, temperature and self heating effects [96]. Hence BTI and HCI continues to remain one of the major reliability concerns even with equivalent scaling [97]. A concept of design for reliability (DFR) targets to implement intrinsically robust and self-healing circuits using innovative design techniques [28]. Using accurate transistor aging models, circuit reliability analysis methodology and novel design strategies, this design equivalent scaling can be achieved. In this investigation an intuitive analytical methodology to evaluate the contribution of different aging mechanisms to performance degradation of linear circuits was introduced. And few novel design strategies to compensate aging degradation were demonstrated using both analytical and experimental results.

By incorporating innovative process improvements and design strategies the growth

Outlook

of semiconductor industry despite of enhanced degradation mechanisms looks optimistic but the future will show...

List of Symbols and Abbreviations

κ	dielectric constant
μ	charge-carrier effective mobility
$\Sigma\Delta$	Sigma delta
τ	switching time constant
A_o	Open loop amplifier DC gain
C_L	Load Capacitance
C_{ox}	gate oxide capacitance per unit area
CLK	clock signal
cm	common mode
f_c	chopping clock frequency
f_{osc}	oscillator switching frequency
F_{ox}	Vertical Gate Oxide Field
I_{DD}	supply current
I_d	Drain current
k	Boltzmann constant
L	transistor gate length
N	number of stages
R_s	Switch equivalent resistance
R_{on}	ON resistance
T	Temperature
T_{inv}	Electrically measured oxide thickness

List of Symbols and Abbreviations

t_{ox}	thickness of gate oxide
V_{cm}	Common mode voltage
V_{DD}	supply voltage
V_{ds}	Drain to source voltage
V_{FS}	full scale voltage
V_{gs}	Gate to source voltage
V_{is}	Input referred offset
V_{od}	Gate to source overdrive voltage
V_{os}	Output referred offset
V_{p-p}	peak-to-peak voltage
V_{SS}	Ground
V_{th}	threshold voltage
$Vref$	reference voltage
AC Stress	dynamic stress
ADC	analog to digital converter
Ain	analog input
AMS	analog and mixed signal
AZ	Auto Zeroing
BTI	bias temperature instability
CCCS	current controlled current source
CHCI	conducting hot carrier injection
CHS	Chopper Stabilization
CMOS	Complementary Metal Oxide Semiconductor
DAC	digital to analog converter
DC Stress	static stress
DFR	design for reliability
DNL	differential non-linearity

DSP	Digital Signal Processor
ENOB	effective number of bits
EOC	end of conversion
EoL	end-of-life
EOT	equivalent oxide thickness
FinFET	Fin-shaped field effect transistor
HK	high-κ
IC	Integrated Circuit
IEEE	Institute of Electrical and Electronics Engineers
INL	integral non-linearity
ITRS	International Technology Roadmap for Semiconductors
MASH	multi-stage-noise-shaping
MG	metal-gate
MOSFET	Metal oxide semiconductor field effect transistor
MuGFET	multiple gate field-effect transistor
NBTI	negative bias temperature instability
NCHCI	non-conducting hot carrier injection
nMOSFET	n-channel MOSFET
NTF	noise transfer function
OSR	oversampling ratio
OTA	operational transconductance amplifier
PBTI	positive bias temperature instability
pMOSFET	p-channel MOSFET
PSD	power spectral density
PVT	process, voltage and temperature
RF	Radio Frequency
SAR	Successive approximation register

SC	switched capacitor	
SNDR	signal to noise and distortion ratio	
SOC	start of conversion	
SOI	silicon-on-insulator	
Tr	Transmission gate	
VCO	voltage controlled oscillator	

References

[1] M. Latif, N. Ali, and F. Hussin, "A Case Study for Reliability-Aware in SoC Analog Circuit Design," in *International Conference on Intelligent and Advanced Systems (ICIAS)*, June 2010, pp. 1–6.

[2] R. Thewes, R. Brederlow, C. Schlunder, P. Wieczorek, A. B., A. Hesener, J. Holz, S. Kessel, , and W. Weber, "MOS Transistor Reliability under Analog Operation," *Microelectronics Reliability*, vol. 40, pp. 1545–1554, Aug-Oct 2000.

[3] F. Chouard, M. Fulde, and D. Schmitt-Landsiedel, "Impact of Degradation Mechanisms on Analog Differential Amplifiers," in *IEEE European Solid-State Circuits Conference Fringe Session (ESSCIRC Fringe)*, Sept. 2009.

[4] A. Gines, E. Peralias, and A. Rueda, "A Survey on Digital Background Calibration of ADCs," in *European Conference on Circuit Theory and Design (ECCTD)*, 2009, pp. 101–104.

[5] F. Chouard, S. More, M. Fulde, and D. Schmitt-Landsiedel, "An Analog Perspective on Device Reliability in 32nm High-k Metal Gate Technology," in *14th International Symposium on Design and Diagnostics of Electronic Circuits Systems (DDECS)*, April 2011, pp. 65–70.

[6] L. Lewyn, T. Ytterdal, C. Wulff, and K. Martin, "Analog Circuit Design in Nanoscale CMOS Technologies," *Proceedings of the IEEE*, vol. 97, no. 10, pp. 1687–1714, Oct. 2009.

[7] L. Lewyn, "Physical Design and Reliability Issues in Nanoscale Analog CMOS Technologies," in *NORCHIP*, Nov. 2009, pp. 1–10.

[8] M. Agostinelli, S. Lau, S. Pae, P. Marzolf, H. Muthali, and S. Jacobs, "PMOS NBTI-Induced Circuit Mismatch in Advanced Technologies," in *42nd Annual IEEE International Reliability Physics Symposium Proceedings (IRPS)*, April 2004, pp. 171–175.

[9] N. Jha, P. Reddy, D. Sharma, and V. Rao, "NBTI Degradation and Its Impact for Analog Circuit Reliability," *IEEE Transactions on Electron Devices*, vol. 52, no. 12, pp. 2609–2615, Dec. 2005.

[10] X. Wu, Z. Chen, and P. Madhani, "Physics and Modeling of Transistor Matching Degradation under Matched External Stress," in *IEEE International Conference on Microelectronic Test Structures (ICMTS)*, March 2008, pp. 233–237.

[11] S. Rauch, "Review and Reexamination of Reliability Effects Related to NBTI-Induced Statistical Variations," *IEEE Transactions on Device and Materials Reliability*, vol. 7, no. 4, pp. 524–530, Dec. 2007.

[12] P. Chaparala, D. Brisbin, and J. Shibley, "NBTI in Dual Gate Oxide PMOSFETs," in *8th International Symposium on Plasma- and Process-Induced Damage*, April 2003, pp. 138–141.

[13] S. Rauch, "The Statistics of NBTI-induced V_T and β Mismatch Shifts in pMOSFETs," *IEEE Transactions on Device and Materials Reliability*, vol. 2, no. 4, pp. 89–93, Dec. 2002.

[14] Y. Chen, J. Zhou, S. Tedja, F. Hui, and A. Oates, "Stress-Induced MOSFET Mismatch for Analog Circuits," in *IEEE International Integrated Reliability Workshop Final Report (IRW)*, 2001, pp. 41–43.

[15] P. Ferreira, H. Petit, and J.-F. Naviner, "CMOS 65 nm Wideband LNA Reliability Estimation," in *Joint IEEE North-East Workshop on Circuits and Systems and TAISA Conference (NEWCAS-TAISA)*, July 2009, pp. 1–4.

[16] W. Wang, V. Reddy, A. Krishnan, R. Vattikonda, S. Krishnan, and Y. Cao, "Compact Modeling and Simulation of Circuit Reliability for 65-nm CMOS Technology," *IEEE Transactions on Device and Materials Reliability*, vol. 7, no. 4, pp. 509–517, Dec. 2007.

[17] V. Huard, C. Parthasarathy, A. Bravaix, T. Hugel, C. Guerin, and E. Vincent, "Design-in-Reliability Approach for NBTI and Hot-Carrier Degradations in Advanced Nodes," *IEEE Transactions on Device and Materials Reliability*, vol. 7, no. 4, pp. 558–570, Dec. 2007.

[18] M. Ruberto, T. Maimon, Y. Shemesh, A. Desormeaux, W. Zhang, and C.-S. Yeh, "Consideration of Age Degradation in the RF Performance of CMOS Radio Chips for High Volume Manufacturing," in *IEEE Digest of Papers from Radio Frequency Integrated Circuits (RFIC) Symposium*, June 2005, pp. 549–552.

[19] C. Schlunder, R. Brederlow, B. Ankele, A. Lill, K. Goser, and R. Thewes, "On the Degradation of p-MOSFETs in Analog and RF Circuits under Inhomogeneous Negative Bias Temperature Stress," in *41st Annual IEEE International Reliability Physics Symposium Proceedings (IRPS)*, March 2003, pp. 5–10.

[20] A. Krishnan, V. Reddy, S. Chakravarthi, J. Rodriguez, S. John, and S. Krishnan, "NBTI Impact on Transistor and Circuit: Models, Mechanisms and Scaling Effects [MOSFETs]," in *IEEE International Technical Digest on Electron Devices Meeting (IEDM)*, Dec. 2003, pp. 14.5.1–14.5.4.

[21] J. Lin, S. Chen, H. Chen, H. Lin, Z. Jhou, S. Chou, J. Ko, T. Lei, and H. Haung, "Matching Variation after HCI Stress in Advanced CMOS Technology for Analog Applications," in *IEEE International Integrated Reliability Workshop Final Report (IRW)*, Oct. 2005, pp. 107–110.

[22] J. Martin-Martinez, B. Kaczer, J. Boix, N. Ayala, R. Rodriguez, M. Nafria, X. Aymerich, P. Zuber, B. Dierickx, and G. Groeseneken, "Circuit-Design Oriented Modelling of the Recovery BTI Component and Post-BD Gate Currents," in *Spanish Conference on Electron Devices (CDE)*, Feb. 2009, pp. 156–159.

[23] F. Chouard, M. Fulde, and D. Schmitt-Landsiedel, "An Aging Suppression and Calibration Approach for Differential Amplifiers in Advanced CMOS Technologies," in *IEEE European Solid-State Circuits Conference Proceedings (ESSCIRC)*, Sept. 2011, pp. 251–254.

[24] A. Ghosh, R. Franklin, and R. Brown, "Analog Circuit Design Methodologies to Improve Negative-Bias Temperature Instability Degradation," in *23rd International Conference on VLSI Design (VLSID)*, Jan. 2010, pp. 369–374.

[25] A. Kawasumi, Y. Takeyama, O. Hirabayashi, K. Kushida, Y. Fujimura, and T. Yabe, "A Low-Supply-Voltage-Operation SRAM With HCI Trimmed Sense Amplifiers," *IEEE Journal of Solid-State Circuits*, vol. 45, no. 11, pp. 2341–2347, Nov. 2010.

[26] P. Mak and R. Martins, "High-Mixed-Voltage RF and Analog CMOS Circuits Come of Age," *IEEE Circuits and Systems Magazine*, vol. 10, no. 4, pp. 27–39, Fourth Quarter, 2010.

[27] R. Carlsten, J. Ralston-Good, and D. Goodman, "An Approach to Detect Negative Bias Temperature Instability (NBTI) in Ultra-Deep Submicron Technologies," in *IEEE International Symposium on Circuits and Systems (ISCAS)*, May 2007, pp. 1257–1260.

[28] E. Maricau and G. Gielen, "Computer-Aided Analog Circuit Design for Reliability in Nanometer CMOS," *IEEE Journal on Emerging and Selected Topics in Circuits and Systems*, vol. 1, no. 1, pp. 50–58, March 2011.

[29] B. Tudor, J. Wang, Z. Chen, R. Tan, W. Liu, and F. Lee, "An Accurate and Scalable MOSFET Aging Model for Circuit Simulation," in *12th International Symposium on Quality Electronic Design (ISQED)*, March 2011, pp. 448–451.

[30] X. Pan and H. Graeb, "Reliability Analysis of Analog Circuits using Quadratic Lifetime Worst-Case Distance Prediction," in *IEEE Custom Integrated Circuits Conference (CICC)*, Sept. 2010, pp. 1–4.

[31] G. Gielen, E. Maricau, and P. De Wit, "Design Automation towards Reliable Analog Integrated Circuits," in *IEEE/ACM International Conference on Computer-Aided Design (ICCAD)*, Nov. 2010, pp. 248–251.

[32] E. Maricau and G. Gielen, "Efficient Variability-Aware NBTI and Hot Carrier Circuit Reliability Analysis," *IEEE Transactions on Computer-Aided Design of Integrated Circuits and Systems*, vol. 29, no. 12, pp. 1884–1893, Dec. 2010.

[33] J. Martin-Martinez, R. Rodriguez, M. Nafria, and X. Aymerich, "Time-Dependent Variability Related to BTI Effects in MOSFETs: Impact on CMOS Differential Amplifiers," *IEEE Transactions on Device and Materials Reliability (TDMR)*, vol. 9, no. 2, pp. 305–310, June 2009.

[34] F. Chouard, M. Fulde, C. Werner, and D. Schmitt-Landsiedel, "A Test Concept for Circuit Level Aging Demonstrated by a Differential Amplifier," in *IEEE International Reliability Physics Symposium (IRPS)*, April 2010, pp. 826–829.

[35] F. Chouard, M. Fulde, and D. Schmitt-Landsiedel, "Reliability Assessment of Voltage Controlled Oscillators in 32nm High-k Metal Gate Technology," in *IEEE European Solid-State Circuits Conference Proceedings (ESSCIRC)*, Sept. 2010, pp. 410–413.

[36] B. Yan, J. Qin, J. Dai, Q. Fan, and J. Bernstein, "Reliability Simulation and Design Consideration of High Speed ADC Circuits," in *IEEE International Integrated Reliability Workshop Final Report (IRW)*, Oct. 2008, pp. 125–128.

[37] X. Chen, S. Samavedam, V. Narayanan, K. Stein, C. Hobbs, C. Baiocco, W. Li, D. Jaeger, M. Zaleski, H. Yang, N. Kim, Y. Lee, D. Zhang, L. Kang, J. Chen, H. Zhuang, A. Sheikh, J. Wallner, M. Aquilino, J. Han, Z. Jin, J. Li, G. Massey, S. Kalpat, R. Jha, N. Moumen, R. Mo, S. Kirshnan, X. Wang, M. Chudzik, M. Chowdhury, D. Nair, C. Reddy, Y. Teh, C. Kothandaraman, D. Coolbaugh, S. Pandey, D. Tekleab, A. Thean, M. Sherony, C. Lage, J. Sudijono, R. Lindsay, J. Ku, M. Khare, and A. Steegen, "A Cost Effective 32nm High-k Metal Gate CMOS Technology for Low Power Applications with Single-metal Gate-first Process," in *Symposium on VLSI Technology*, 17-19 June 2008, pp. 88–89.

[38] G. Ribes, J. Mitard, M. Denais, S. Bruyere, F. Monsieur, C. Parthasarathy, E. Vincent, and G. Ghibaudo, "Review on High-k Dielectrics Reliability Issues," *IEEE Transactions on Device and Materials Reliability*, vol. 5, no. 1, pp. 5 – 19, March 2005.

[39] K. T. Lee, C. Y. Kang, O. S. Yoo, R. Choi, B. H. Lee, J. Lee, H.-D. Lee, and Y.-H. Jeong, "PBTI-Associated High-Temperature Hot Carrier Degradation of nMOSFETs with Metal-Gate/High- k Dielectrics," *IEEE Electron Device Letters*, vol. 29, no. 4, pp. 389–391, April 2008.

[40] A. W. Strong, E. Y. Wu, R.-P. Vollertsen, J. Sune, G. L. Rosa, T. D. Sullivan, and S. E. I. Rauch, *Reliability Wearout Mechanisms in Advanced CMOS Technologies*. Wiley-IEEE Press, 2009.

[41] R. Entner, "Modeling and Simulation of Negative Bias Temperature Instability," Ph.D. dissertation, Technischen Universität Wien, April 2007.

[42] Y. Miura and Y. Matukura, "Investigation of Silicon-Silicon Dioxide Interface Using MOS Structure," in *Japanese Journal of Applied Physics*, vol. 5, 1966, p. 180.

[43] K. Zhao, J. Stathis, B. Linder, E. Cartier, and A. Kerber, "PBTI under Dynamic Stress: From a Single Defect Point of View," in *IEEE International Reliability Physics Symposium (IRPS)*, April 2011, pp. 4A.3.1–4A.3.9.

[44] T. Grasser, T. Aichinger, G. Pobegen, H. Reisinger, P. Wagner, J. Franco, M. Nelhiebel, and B. Kaczer, "The Permanent Component of NBTI: Composition and Annealing," in *IEEE International Reliability Physics Symposium (IRPS)*, April 2011, pp. 6A.2.1–6A.2.9.

[45] H. Reisinger, T. Grasser, K. Ermisch, H. Nielen, W. Gustin, and C. Schlunder, "Understanding and Modeling AC BTI," in *IEEE International Reliability Physics Symposium (IRPS)*, April 2011, pp. 6A.1.1–6A.1.8.

[46] S. Ramey, C. Prasad, M. Agostinelli, S. Pae, S. Walstra, S. Gupta, and J. Hicks, "Frequency and Recovery Effects in High-k; BTI Degradation," in *IEEE International Reliability Physics Symposium (IRPS)*, 2009, pp. 1023–1027.

[47] T. Grasser, B. Kaczer, P. Hehenberger, W. Gos, R. O'Connor, H. Reisinger, W. Gustin, and C. Schlunder, "Simultaneous Extraction of Recoverable and Permanent Components Contributing to Bias-Temperature Instability," in *IEEE International Electron Devices Meeting (IEDM)*, 10-12 Dec. 2007, pp. 801–804.

[48] B. Kaczer, T. Grasser, J. Martin-Martinez, E. Simoen, M. Aoulaiche, P. Roussel, and G. Groeseneken, "NBTI from the Perspective of Defect States with Widely Distributed Time Scales," in *IEEE International Reliability Physics Symposium (IRPS)*, 26-30 April 2009, pp. 55–60.

[49] H. Reisinger, T. Grasser, K. Hofmann, W. Gustin, and C. Schlunder, "The Impact of Recovery on BTI Reliability Assessments," in *IEEE International Integrated Reliability Workshop Final Report (IRW)*, Oct. 2010, pp. 12–16.

[50] H. Reisinger, "NBTI: Recent Findings and Controversial Topics," in *IEEE International Reliability Physics Symposium (IRPS)*, April 2011.

[51] T. Nigam and E. Harris, "Lifetime Enhancement under High Frequency NBTI Measured on Ring Oscillators," in *44th Annual IEEE International Reliability Physics Symposium Proceedings (IRPS)*, March 2006, pp. 289–293.

[52] G. Chen, K. Chuah, M. Li, D. Chan, C. Ang, J. Zheng, Y. Jin, and D. Kwong, "Dynamic NBTI of PMOS Transistors and its Impact on Device Lifetime," in *41st Annual IEEE International Reliability Physics Symposium Proceedings (IRPS)*, 4 April 2003, pp. 196–202.

[53] W. Abadeer and W. Ellis, "Behavior of NBTI under AC Dynamic Circuit Conditions," in *41st Annual IEEE International Reliability Physics Symposium Proceedings (IRPS)*, March 2003, pp. 17–22.

[54] V. Huard, C. Parthasarathy, A. Bravaix, C. Guerin, and E. Pion, "CMOS Device Design-in Reliability Approach in Advanced Nodes," in *IEEE International Reliability Physics Symposium (IRPS)*, 26-30 April 2009, pp. 624–633.

[55] H. Reisinger, T. Grasser, W. Gustin, and C. Schlunder, "The Statistical Analysis of Individual Defects Constituting NBTI and its Implications for Modeling DC- and AC-stress," in *IEEE International Reliability Physics Symposium (IRPS)*, May 2010, pp. 7 –15.

[56] T. Grasser, H. Reisinger, W. Goes, T. Aichinger, P. Hehenberger, P.-J. Wagner, M. Nelhiebel, J. Franco, and B. Kaczer, "Switching Oxide Traps as the Missing Link between Negative Bias Temperature Instability and Random Telegraph Noise," in *IEEE International Electron Devices Meeting (IEDM)*, Dec. 2009, pp. 1 –4.

[57] A. Bravaix, C. Guerin, V. Huard, D. Roy, J. Roux, and E. Vincent, "Hot-Carrier Acceleration Factors for Low Power Management in DC-AC Stressed 40nm NMOS node at High Temperature," in *IEEE International Reliability Physics Symposium*, 2009, pp. 531 –548.

[58] A. Bravaix, C. Guerin, D. Goguenheim, V. Huard, D. Roy, C. Besset, S. Renard, Y. Randriamihaja, and E. Vincent, "Off State Incorporation into the 3 energy mode Device Lifetime Modeling for Advanced 40nm CMOS node," in *IEEE International Reliability Physics Symposium (IRPS)*, May 2010, pp. 55 –64.

[59] D. Varghese, M. Alam, and B. Weir, "A Generalized, IB-independent, Physical HCI Lifetime Projection Methodology based on Universality of Hot-Carrier Degradation," in *IEEE International Reliability Physics Symposium (IRPS)*, May 2010, pp. 1091 –1094.

[60] RelXpert, *Users Manuals BSIMPro+/RelXpert/UltraSim*, Cadence Design Systems Inc, 2010.

[61] S. More, M. Fulde, F. Chouard, and D. Schmitt-Landsiedel, "Sensitivity Analysis Based Analytical Evaluation of Aging Degradation in Linear Circuits," in *IEEE European Solid-State Circuits Conference Fringe Session (ESSCIRC Fringe)*, 13-17 Sept. 2010.

[62] *Spectre Circuit Simulator Reference*, Cadence Design Systems, July 2002.

[63] S. More, M. Fulde, F. Chouard, and D. Schmitt-Landsiedel, "Reliability Analysis of Buffer Stage in Mixed Signal Application," in *Advances in Radio Science Journal (ARS)*, vol. 9, 2011, pp. 225–230.

[64] H. Mostafa, M. Anis, and M. Elmasry, "Adaptive Body Bias for Reducing the Impacts of NBTI and Process Variations on 6T SRAM Cells," *IEEE Transactions on Circuits and Systems I: Regular Papers*, no. 99, 2011.

[65] S. More, M. Fulde, F. Chouard, and D. Schmitt Landsiedel, "Reducing Impact of Degradation on Analog Circuits by Chopper Stabilization and Autozeroing," in *12th International Symposium on Quality Electronic Design (ISQED)*, March 2011, pp. 8–13.

[66] C. Enz and G. Temes, "Circuit Techniques for Reducing the Effects of Op-Amp Imperfections: Autozeroing, Correlated Double Sampling and Chopper Stabilization," *Proceedings of the IEEE*, vol. 84, no. 11, pp. 1584–1614, Nov. 1996.

[67] K.-C. Hsieh and P. Gray, "A Low-Noise Chopper-Stabilized Differential Switched-Capacitor Filtering Technique," in *IEEE International Solid-State Circuits Conference (ISSCC)*, vol. 24, Feb. 1981, pp. 128–129.

[68] P. E. Allen and D. R. Holberg, *CMOS Analog Circuit Design*, 2nd ed. Oxford University Press, 2002.

[69] M. Liu, *Demystifying Switched Capacitor Circuits*, 1st ed. Newnes, 2006.

[70] J. Keane, X. Wang, D. Persaud, and C. Kim, "An All-In-One Silicon Odometer for Separately Monitoring HCI, BTI, and TDDB," *IEEE Journal of Solid-State Circuits*, vol. 45, no. 4, pp. 817–829, April 2010.

[71] V. Reddy, A. Krishnan, A. Marshall, J. Rodriguez, S. Natarajan, T. Rost, and S. Krishnan, "Impact of Negative Bias Temperature Instability on Digital Circuit Reliability," in *40th Annual IEEE International Reliability Physics Symposium Proceedings (IRPS)*, 2002, pp. 248–254.

[72] T. Nigam, "Impact of Transistor Level Degradation on Product Reliability," in *IEEE Custom Integrated Circuits Conference (CICC)*, Sept. 2009, pp. 431–438.

[73] B. Razavi, *Design of Analog CMOS Integrated Circuits*, 1st ed. McGraw-Hill, 2000.

[74] V. Huard, M. Denais, and C. Parthasarathy, "NBTI Degradation: From Physical Mechanisms to Modelling," *Microelectronics and Reliability*, vol. 46, no. 1, pp. 1–23, 2006.

[75] S. Zhu, A. Nakajima, T. Ohashi, and H. Miyake, "Enhancement of BTI Degradation in pMOSFETs under High-frequency Bipolar Gate Bias," *IEEE Electron Device Letters*, vol. 26, no. 6, pp. 387–389, June 2005.

[76] Y. Liang, Z. Wu, and B. Li, "A New 12-Bit Fully Differential SAR ADC for Wireless Implantable Neural Recording System," in *IEEE International Conference of Electron Devices and Solid-State Circuits (EDSSC)*, Dec. 2009, pp. 399–402.

[77] M. Fulde, D. Schmitt-Landsiedel, and G. Knoblinger, "Transient Variations in Emerging SOI Technologies: Modeling and Impact on Analog/Mixed-Signal Circuits," in *IEEE International Symposium on Circuits and Systems (ISCAS)*, 27-30 May 2007, pp. 1249–1252.

[78] J. A. McNeill, K. Y. Chan, M. C. W. Coln, C. L. David, and C. Brenneman, "All-Digital Background Calibration of a Successive Approximation ADC using the Split ADC Architecture," *IEEE Transactions on Circuits and Systems I: Regular Papers*, vol. 58, no. 10, pp. 2355–2365, 2011.

[79] M. Kinyua, F. Maloberti, and W. Gosney, "Digital Background Auto-calibration of DAC Non-Linearity in Pipelined ADCs," in *Proceedings of the International Symposium on Circuits and Systems (ISCAS)*, vol. 1, May 2004, pp. I-13–I-16.

[80] E. Siragusa and I. Galton, "Gain Error Correction Technique for Pipelined Analogue-to-Digital Converters," *Electronics Letters*, vol. 36, no. 7, pp. 617–618, Mar. 2000.

[81] S. Schwarz, "Modeling of Self Correcting Pipeline A/D Converter using Redundancy and Self Calibration Techniques," Bachelor's Thesis, Technische Universität München, Jan. 2010.

[82] H. Wang, X. Wang, P. Hurst, and S. Lewis, "Nested Digital Background Calibration of a 12-bit Pipelined ADC Without an Input SHA," *IEEE Journal of Solid-State Circuits*, vol. 44, no. 10, pp. 2780–2789, Oct. 2009.

[83] B. Murmann and B. Boser, "A 12-bit 75-MS/s Pipelined ADC using Open-Loop Residue Amplification," *IEEE Journal of Solid-State Circuits*, vol. 38, no. 12, pp. 2040–2050, Dec. 2003.

[84] Y.-S. Shu and B.-S. Song, "A 15-bit Linear 20-ms/s Pipelined ADC Digitally Calibrated With Signal-Dependent Dithering," *IEEE Journal of Solid-State Circuits*, vol. 43, no. 2, pp. 342–350, Feb. 2008.

[85] Y. Geerts, M. Steyaert, and W. Sansen, *Design of Multi-Bit Delta-Sigma A/D Converters*, 1st ed. Springer, 2002.

[86] Q. Yang, "Modeling of Multi-bit Sigma Delta ADC using Self-calibration Techniques," Bachelor's Thesis, Technische Universität München, April 2010.

[87] J. B. da Silva, "High-Performance Delta-Sigma Analog-to-Digital Converters," Ph.D. dissertation, Oregon State University, July 2004.

[88] M. Ruf, "Impact of Degradation Mechanisms on Sigma-Delta ADC Performance And Implementation of Correction Technique," Bachelor's Thesis, Technische Universität München, August 2011.

[89] P. Kiss, "Adaptive Digital Compensation of Analog Circuit Imperfections for Cascaded Delta-Sigma Analog-to-Digital Converters," Oregon State University, Tech. Rep., August 1999.

[90] X. Wang, "A Fully Digital Technique for the Estimation and Correction of the DAC Error in Multi-bit Delta Sigma ADCs," Ph.D. dissertation, Oregon State University, June 2004.

[91] G. Gielen, E. Maricau, and P. De Wit, "Analog Circuit Reliability in Sub-32 Nanometer CMOS: Analysis and Mitigation," in *Design, Automation Test in Europe Conference Exhibition (DATE)*, March 2011, pp. 1–6.

[92] T. Grasser, B. Kaczer, W. Goes, H. Reisinger, T. Aichinger, P. Hehenberger, P.-J. Wagner, F. Schanovsky, J. Franco, P. Roussel, and M. Nelhiebel, "Recent Advances in Understanding the Bias Temperature Instability," in *IEEE International Electron Devices Meeting (IEDM)*, Dec. 2010, pp. 4.4.1–4.4.4.

[93] M. Fulde, *Variation Aware Analog and Mixed-Signal Circuit Design in Emerging Multi-Gate CMOS Technologies*, ser. Advanced Microelectronics. Springer, 2010, vol. 28.

[94] International Technology Roadmap for Semiconductors (ITRS), "International Technology Roadmap for Semiconductors, 2009 Edition," Tech. Rep., 2009.

[95] G. Groeseneken, F. Crupi, A. Shickova, S. Thijs, D. Linten, B. Kaczer, N. Collaert, and M. Jurczak, "Reliability Issues in MuGFET Nanodevices," in *IEEE International Reliability Physics Symposium (IRPS)*, May 2008, pp. 52–60.

[96] Y. Wang, S. Cotofana, and L. Fang, "A Unified Aging Model of NBTI and HCI Degradation towards Lifetime Reliability Management for Nanoscale MOSFET Circuits," in *IEEE/ACM International Symposium on Nanoscale Architectures (NANOARCH)*, June 2011, pp. 175–180.

[97] C.-W. Lee, I. Ferain, A. Afzalian, K.-Y. Byun, R. Yan, N. Dehdashti, P. Razavi, W. Xiong, J. Colinge, C. Colinge, and D. Ioannou, "Hot Carrier (HC) and Bias-Temperature-Instability (BTI) Degradation of MuGFETs on Silicon Oxide and Silicon Nitride Buried Layers," in *Proceedings of the IEEE European Solid State Device Research Conference (ESSDERC)*, Sept. 2009, pp. 261–264.

i want morebooks!

Buy your books fast and straightforward online - at one of world's fastest growing online book stores! Environmentally sound due to Print-on-Demand technologies.

Buy your books online at
www.get-morebooks.com

Kaufen Sie Ihre Bücher schnell und unkompliziert online – auf einer der am schnellsten wachsenden Buchhandelsplattformen weltweit! Dank Print-On-Demand umwelt- und ressourcenschonend produziert.

Bücher schneller online kaufen
www.morebooks.de

   VDM Verlagsservicegesellschaft mbH
Heinrich-Böcking-Str. 6-8   Telefon: +49 681 3720 174   info@vdm-vsg.de
D - 66121 Saarbrücken   Telefax: +49 681 3720 1749   www.vdm-vsg.de

Printed by Books on Demand GmbH, Norderstedt / Germany